CSE EXAM
SECRETS

THEMATICS TO A*

GCSE Exam Secrets

Mathematics to A*

Ma[illegible] & Brian Seager

CONTENTS

THIS BOOK AND YOUR GCSE EXAMS

Introduction

This book is designed to help you get better results.

- Look at the grade A and C candidates' answers and see if you could have done better.
- Try the exam practice questions and then look at the answers.
- Make sure you understand why the answers given are correct.
- When you feel ready, try the GCSE mock exam papers.

If you perform well on the questions in this book you should do well in the examination. Remember that success in examinations is about hard work, not luck.

What examiners look for

The obvious:

- Work which is legible, set out clearly, and easy to follow and understand; that you have used pen, not pencil, except in drawings, and that you have used the appropriate equipment.
- That drawings and graphs are neat, and graphs are labelled.
- That you always indicate how you obtain your answers.
- The right answer!

Exam technique

- Start with question 1 and work through the paper in order.
- If you cannot answer a question, leave it out and return to it at the end.
- For explanation questions, use the mark allocation to guide you on how many points to make.
- Make sure you answer the question that has been asked.
- Set your work out neatly and logically. Untidy work is not only more difficult for the examiner to read but you could also misread your own figures.
- Show all the necessary working so that you can earn the method marks even if you make a numerical slip. You need to convince the examiner that you know what you are doing.
- Do not be sloppy with algebraic notation and manipulation, brackets and negatives in particular.
- Do rough calculations to check your answers and make sure they are reasonable.
- Do not plan to have time left over at the end. If you do, use it usefully. Check you have answered all the questions, check arithmetic and read longer answers to make sure you have not made silly mistakes or missed things out.

See also Top Tips for Exam Success on page 7.

DIFFERENT TYPES OF EXAM QUESTIONS

Understand the question

It is important that you understand what the examiner means when using words like 'State', 'Find' and 'Deduce'? Here is a brief glossary to help you:

Write down, state – no working out or explanation needed.
Calculate, find, show, solve – some working out needed. Include enough working to make your method clear.
Deduce, hence – make use of an earlier answer to establish the result.
Prove – set out a concise logical argument, making the reasons clear.
Sketch – show the general shape of a graph, its relationship with the axes and points of special significance.
Draw – if a graph, plot accurately using graph paper; use geometrical instruments carefully for other diagrams.
Find the exact value – leave in fractions, surds or π. Rounded results from a calculator will not earn the marks.

Multistep questions

Multistep questions require you to obtain the answer to a problem where more than one step is required. Questions which are not multistep are sometimes called **structured**. In these, there may still be more than one step, but each stage is set out in the question. Here is an example.

Structured questions

The diagram shows one end of a railway van.
The roof SQ is a circular arc, centre B, the mid-point of RP.
RP = 2.10 m, PQ = 2.33 m.
Calculate:

(a) the length of BQ [2]
(b) the angle BQP [3]
(c) the length of the arc QS. [3]

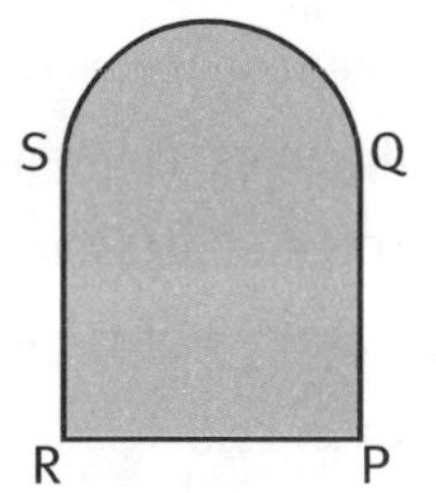

Multistep question

Calculate the length of the arc QS. [8]

In the second form of the question, it is necessary to work out the radius of the circle (BQ) and the angle SBQ (twice angle BQP) before finding the length of the arc. The work is exactly the same but in the second case, you are not told what the intermediate steps are. You need to work out the strategy for yourself.

Using a calculator

Half of GCSE Mathematics papers must be answered without the use of calculating aids. This means that either the first paper is non-calculator, or the first section of each paper is non-calculator.
In this book, questions which have to be answered without a calculator are marked with this symbol.
For the mock examination at the end of the book, the first paper is non-calculator.

Algebra

In a Higher Tier GCSE paper, about one-third of the marks will be for algebra. Most of these will be for manipulative algebra, that is expanding brackets, simplifying, factorising, solving equations and inequalities, changing the subject of a formula. Some of the algebra marks will be for graphs, substitution and sequences.

WHAT IS NEEDED TO BE AWARDED A GRADE A IN MATHEMATICS

The following statements are a list of the minimum requirements of those aspects of mathematics that you are expected to know for the award of a grade A on the written examination papers. They are based on the generic descriptors that are used by all examination boards and do not include any requirements that would be covered by coursework.

An grade A candidate should be able to:

- Give reasons, justify and explain solutions to problems.
- Use mathematical language and symbols in reasoned argument.
- Understand rational and irrational numbers.
- Determine bounds of intervals.
- Understand and use direct and inverse proportion.
- Rearrange and manipulate formulae; solve equations; find factors; understand rules of indices; solve simultaneous equations algebraically and graphically; solve other problems graphically.
- Use sine, cosine and tangent of angles of any size.
- Understand and use Pythagoras' theorem in 2 and 3 dimensions.
- Use conditions for congruent triangles in formal geometric proofs.
- Calculate lengths of arcs and areas of sectors.
- Calculate the surface area of cylinders and the volumes of cones and spheres.
- Construct and interpret histograms.
- Understand different methods of sampling.
- Recognise when and how to use probabilities associated with independent and mutually exclusive events.

What do I need for a grade A*?

Unlike grade A, there are no criteria for the A* grade. You need to get nearly all of the questions right. You should have no weaknesses in any area of the specification. The boundary mark is determined to be the same number of marks above grade A as grade B is below grade A.

HOW TO BOOST YOUR GRADE

Common areas of difficulty

These are the parts of the Higher tier specification which candidates find difficult. You will find help in revising these in *Letts Revise GCSE Mathematics Study Guide*. The references are given in brackets.

- Algebra, especially solving equations, inequalities, changing the subject of a formula, algebraic fractions (2.2, 2.7, 2.8, 2.9, 2.10)
- Proportion, particularly inverse proportion (2.6)
- Reversed percentages, e.g. finding cost before VAT was added (1.5)
- Manipulating surds (1.8)
- Solving 3-D problems using Pythagoras and trigonometry (3.3, 3.4)
- Histograms and frequency density (4.3)
- Conditional probability, where the probability of an event depends on what has happened before (4.4)
- Vectors (3.9)

Top tips for exam success

- Practise all aspects of manipulative algebra, solving equations, rearranging formulae, expanding brackets, factorising, simplifying.
- Practise answering questions without the use of a calculator.
- Practise answering questions with more than one step to the answer (multistep questions), e.g. finding the radius of a sphere with the same volume as a given cone.
- Make your drawings and graphs neat and accurate.
- Practise answering questions that ask for an explanation or proof. Your answers should be concise and use mathematical terms where appropriate.
- Don't forget to check your answers, especially to see that they are reasonable. The mean height of a group of men will not be 187 metres!
- Lay out your working carefully and concisely. Write down the calculations that you are going to make. You usually get marks for showing a correct method.
- Know what is on, and what is not on the formula sheet before the examination.
- Make sure you can use your calculator efficiently. Write down the figures on your calculator and then make suitable rounding. Don't round the numbers during the calculation. This will often result in an inaccurate answer.
- Make sure you have read the question carefully so you give the answer the examiner wants!

CHAPTER 1

Number

To revise this topic more thoroughly, see Chapter 1 in *Letts Revise GCSE Mathematics Study Guide.*

Try this sample GCSE question and then compare your answers with the Grade C and Grade A model answers on the next page. *Do not use a calculator in this question.*

a Find fractions equivalent to these decimals.
Give your answers in their lowest terms.

(i) 0.124

....

.... **[1]**

(ii) $0.\dot{1}2\dot{4}$

....

....

.... **[4]**

b Simplify these, writing each as a simple surd.

(i) $\sqrt{2} \times \sqrt{8}$

.... **[1]**

(ii) $\dfrac{1}{\sqrt{3}}$

.... **[1]**

(iii) $(1 + \sqrt{2})^2$

....

.... **[2]**

(iv) $\dfrac{1}{1 + \sqrt{2}}$

....

.... **[3]**

(Total 12 marks)

These two answers are at grades C and A. Compare which one your answer is closest to and think how you could have improved it.

GRADE C ANSWER

Michael

Correct. → a (i) $0.124 = \frac{124}{1000} = \frac{62}{500} = \frac{31}{250}$ ✓

(ii) $0.\dot{1}2\dot{4} = 0.124124124...$ ← *Michael has multiplied by 10, which is the right idea but × 1000 is what is needed.*

$1.24 = 1.24...$

$= 1.12$ ✗

b (i) $\sqrt{2} \times \sqrt{8} = \sqrt{(2 \times 8)} = \sqrt{16} = 4$ ✓ ← *This is correct and clearly set out.*

No progress towards rationalising the denominator and a classic error, $\frac{1}{3} = 0.3$. → (ii) $\frac{1}{\sqrt{3}} = \sqrt{\frac{1}{3}} = \sqrt{0.3}$ ✗

(iii) $(1 + \sqrt{2})^2 = 1^2 + (\sqrt{2})^2 = 1 + 2 = 3$ ✗ ← *Another common error, remember* $(a + b)^2 = a^2 + 2ab + b^2$

(iv) ?

2 marks = Grade C answer

Grade booster ···> move a C to a B

Michael will struggle to make a C grade making mistakes like this. Other questions will need to be much better than this if he is to succeed. However, much of this question is on work which is higher than C.

GRADE A ANSWER

Leon

Correct. → a (i) $\frac{124}{1000} = \frac{31}{250}$ ✓

(ii) $F = 0.124124...$

$1000F = 124.124124...$ ✓ ← *This work is correct and clear.*

$999F = 124$ ✓

This fraction will not simplify. → $F = \frac{124}{999}$ ✓✓

b (i) $\sqrt{2} \times 2\sqrt{2} = 4$ ✓ ← *Correct again.*

(ii) $\frac{\sqrt{3}}{3}$ ✓ ← *This has clearly been understood.*

Silly to lose a mark here for not adding 2 and 1! → (iii) $1 + 2\sqrt{2} + 2$ ✓

(iv) $\frac{1 + \sqrt{2}}{(1 + \sqrt{2})^2} = ?$ ✗ ← *Almost the right idea but multiplying by $1 - \sqrt{2}$ was required.*

8 marks = Grade A answer

Grade booster ···> move A to A*

Leon clearly understands most of this work and should get a grade A if other questions are as good. However, 2 marks were lost for not following the instructions in the question. With care to avoid such slips, even an A* could be possible.

Try this sample GCSE question and then compare your answers with the Grade C and Grade A model answers on the next page.

1 An ancient city is being excavated by archaeologists. The city was a rectangle 982 m by 463 m.

a What is the upper bound of the area covered by the city?

....

....

....

....

....

.... **[3]**

b Find the lower bound of the perimeter.

....

....

....

....

....

.... **[2]**

c The population of the city was estimated to have been 100 000, to the nearest thousand.
Find the lower bound of the population density.

....

....

....

....

....

.... **[4]**

(Total 9 marks)

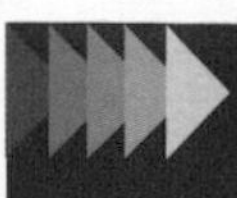

These two answers are at grades C and A. Compare which one your answer is closest to and think how you could have improved it. See Revise Mathematics Guide section 1.5 for further help.

GRADE C ANSWER

Darren

a $982.4 \times 463.4 = 455244$ m² ✓ ← *This is a common error, to use values below the upper bound.*

This is only half the perimeter. → **b** $981.5 + 462.5 = 1444$ m ✓

c $100000 \div 455244 = 0.22$ ✓ ← *Correct method, but no units given*

3 marks = Grade C answer

Grade booster ···> move a C to a B

Darren has not given the upper bounds for the measurements in part (a), thinking he must use numbers less than the bounds. Further marks are lost in part (b) by forgetting to double. At Grade C, he would not be expected to do part (c), although he has gained a method mark for dividing number of people by area.

GRADE A ANSWER

Karen

a $982.5 \times 463.5 = 455388.75$ ✓✓✓ *Completely correct.*

Correct again, although addition could have been replaced by multiplying by 2. → **b** $981.5 + 462.5 + 981.5 + 462.5$
$= 2888$ ✓✓

c $99500 \div (981.5 \times 462.5)$
$= 0.2192$ ✓✓ ← *No units stated.*

7 marks = Grade A answer

Grade booster ···> move A to A*

Karen should get Grade A, if this performance is repeated in other questions. For A*, there is no room for mistakes such as not dividing by the maximum area in part (c) to obtain the minimum density. A further mark is lost for omitting the units (m², m, people/m²).

QUESTION BANK

1 a) Write 24 as a product of prime factors.

....

....

.... ②

b) i) Find the lowest common multiple of 24 and 56.

8 2

.... ①

ii) Find the highest common factor of 24 and 56.

8

.... ①

c) i) Find the highest common factor and the lowest common multiple of 36 and 84.

Highest 36

Lowers 21

.... ④

ii) How did the HCF help you to find the LCM? Explain.

....

.... ①

TOTAL 9

2 The number 450 can be written as $2^a \times 3^b \times 5^c$.
Calculate the values of a, b and c.

....

....

.... ③

TOTAL 3

 ANSWERS ON PAGE 22 ANSWERS ON PAGE 22 ANSWERS ON PAGE 22 ANSWERS ON PAGE 22

3 Simplify

a) i) $25^{\frac{1}{2}}$ ii) 3^{-2} iii) 7^0

(3)

b) Five multiplied by the reciprocal of ten.

(1)

c) $\dfrac{(3^2)^4}{(3^4)^{-\frac{1}{2}}}$

(2)

TOTAL 6

4 Write as simply as possible.

a) $\sqrt{90} - \sqrt{40}$

(2)

b) $\sqrt{90} \div \sqrt{40}$

(1)

c) $\sqrt{90} \times \sqrt{40}$

(1)

TOTAL 4

5 George is visiting the UK from the USA.
He can buy goods to take home without paying VAT.
He sees an antique desk priced at £3290, including VAT at 17.5%.

How much does he pay for it?

(3)

TOTAL 3

6 Work out, giving your answers in standard form, correct to 4 significant figures.

a) $63 \times 365.25 \times 24 \times 3600$

(2)

b) $\dfrac{5.73 \times 10^{12}}{3.98 \times 10^4 \times 8.3 \times 10^{-3}}$

(3)

TOTAL 5

ANSWERS ON PAGE 22 ANSWERS ON PAGE 22 ANSWERS ON PAGE 22 ANSWERS ON PAGE 22

Number

7 a) Use index notation to express 144 as a product of prime factors.

....

....

....

.... (3)

b) Simplify.

i) $\dfrac{2^6 \times 5^4}{10^3}$ (2) ii) $\left(2^8\right)^{\frac{1}{2}}$ (1)

....

....

c) Simplify, writing the answers without indices.

i) $64^{-\frac{1}{3}}$ (1) ii) $25^{-\frac{3}{2}}$ (1) iii) $\dfrac{\left(2^{\frac{1}{2}}\right)^3}{\left(2^{-3}\right)^{\frac{1}{2}}}$ 

....

....

....

TOTAL 9

8 Work out.

a) $2\frac{3}{4} + 1\frac{3}{5}$

....

b) $2\frac{3}{4} \times 1\frac{3}{5}$

....

.... (3)

TOTAL 6

9 Work out.

a) $4\frac{1}{4} - 3\frac{2}{3}$

....

.... (3)

b) $4\frac{1}{4} \div 3\frac{2}{3}$

....

.... (3)

TOTAL 6

ANSWERS ON PAGE 22 ANSWERS ON PAGE 22 ANSWERS ON PAGE 22 ANSWERS ON PAGE 22

10 a) Convert these fractions into decimals.

i) $\frac{3}{16}$ (1) ii) $\frac{4}{15}$ (2)

....

b) Convert these decimals into fractions in their lowest terms.

i) 0.093 75 (2) ii) $0.\dot{3}\dot{6}$ (2)

....

....

....

TOTAL 7

---

11 Two fractions have a mean of $\frac{5}{12}$ and range $\frac{1}{3}$.
Work out the two fractions.

....

....

....

.... (4)

TOTAL 4

---

12 a) Add together $0.1\dot{6}$ and $0.2\dot{6}$, expressing the result as a recurring decimal.

....

.... (2)

b) Show the result is correct by converting the decimals into fractions.

....

.... (3)

TOTAL 5

---

13 These are the internal measurements of a baked bean tin.
Circumference 225 mm, correct to nearest 5 mm.
Height 102 mm, correct to the nearest millimetre.

Find the lower bound of the capacity of the tin.

....

....

.... (4)

TOTAL 4

ANSWERS ON PAGE 22 ANSWERS ON PAGE 22 ANSWERS ON PAGE 22 ANSWERS ON PAGE 22

14 A dishwasher costs £440, including VAT at 17.5%.

What is the cost without VAT?

£440

....

....

....

.... (3)

TOTAL 3

15 A 'Supaball' is dropped on the floor from a height of 2 m. Each time it bounces, it rises to 85% of the height from which it fell.

2 m

a) How high will it rise after the second bounce?

....

.... (2)

b) After how many bounces will it rise to a height of less than 1 m?

....

....

.... (2)

TOTAL 4

16 James invested some money at 4.7% interest for one year. At the end of the year there was £16 437.90.

a) How much had he invested?

....

.... (3)

b) The interest rate falls to 3.95%, compound. If he leaves the money in the account, how many years will it be before he has over £20 000?

....

.... (3)

TOTAL 6

ANSWERS ON PAGE 22 ANSWERS ON PAGE 22 ANSWERS ON PAGE 22 ANSWERS ON PAGE 22

17 Mr and Mrs Bristow borrowed £150 000 to buy a house.
Every month 0.5% interest is added to the amount they owe at the beginning of the month, but they pay back £800.

a) How much do they owe at the end of two months?

....

.... ③

b) After 6 months, how much do they owe?

....

....

.... ②

TOTAL 5

---

18 Three friends bought lottery tickets.
They agreed to share any prizes in the ratio of the number of tickets they bought.

a) One week, Sarah bought 6 tickets, Bronwyn 3 and Gail 5.
They won £53 876.
Work out how much each should get.

....

....

....

....

.... ③

b) Another week, Sarah forgot to buy a ticket. The other two shared a £10 prize.
Bronwyn's share was £9.09.
Work out the ratio of the number of tickets they each bought.

....

....

.... ②

TOTAL 5

ANSWERS ON PAGE 22 ANSWERS ON PAGE 22 ANSWERS ON PAGE 22 ANSWERS ON PAGE 22

Number

19 Simplify these surds, giving your answers in the form $a + b\sqrt{c}$.

a) $(1 + \sqrt{2})(1 - \sqrt{2})$

....

.... ②

b) $\dfrac{\sqrt{18}}{\sqrt{6}}$

....

.... ②

c) $(4 - 3\sqrt{2})^2$

....

.... ③

TOTAL 7

20 Paving blocks measure 100 mm by 200 mm.

My drive has an area of 125 m^2.
The pattern of the paving requires red, blue and grey blocks in the ratio 8 : 4 : 3.

Work out the number of blocks of each colour needed.
Give your answer to a sensible degree of accuracy.

100 mm

200 mm

....

....

....

.... ④

TOTAL 4

21 There are deductions from my pay for tax, national insurance and pension contribution.
These are in the ratio 20 : 7 : 6.

a) Last month the tax was £412.
How much was the pension contribution?

....

.... ③

b) The ratio of my pay after deductions to the total deductions is 2 : 1.
My pay after deductions was £1327.
Find how much national insurance I paid.

....

....

....

.... ③

TOTAL 6

ANSWERS ON PAGE 22 ANSWERS ON PAGE 22 ANSWERS ON PAGE 22 ANSWERS ON PAGE 22

22 In an epidemic, the number of new cases is increasing by 32% a week.
At the start, there are 2000 cases.

Work out how many there will be after 4 weeks.

....

....

....

.... ③

TOTAL 3

23 A block of ice is melting.
Each day its mass is 75% of that on the previous day.
At the start, the mass of ice was 400 kg.

How much ice will there be after 10 days?

....

....

.... ③

TOTAL 3

24 A paving block measures 10.0 cm by 20.0 cm by 6.0 cm, all lengths correct to the nearest millimetre.

6.0 cm
10.0 cm
20.0 cm

Find the maximum volume of the block.

....

....

.... ③

TOTAL 3

25 Jo got into a lift with five other people.
Here is the sign in the lift.

MAX LOAD
6 persons
or 450 kg

On the way up to the twelfth floor,
Jo asked everyone's weight.
Here are the results.

70 60 85 60 95 80

Assuming they all gave their weights correct to the nearest 5 kg and the safety limit on the lift is also correct to the nearest 5 kg, work out whether or not the safety conditions were met. Explain your answer.

....

....

....

.... ⑤

TOTAL 5

ANSWERS ON PAGE 22 ANSWERS ON PAGE 22 ANSWERS ON PAGE 22 ANSWERS ON PAGE 22

 A train covers a measured kilometre (correct to the nearest metre) in 20.8 s (correct to the nearest 0.1 s).

Give the possible limits on its speed in km/h.

....

....

....

.... (4)

TOTAL 4

---

27 Use approximations to estimate the values of the following expressions.

a) $2\pi\sqrt{\dfrac{47.3}{9.81}}$ (3)

b) $\dfrac{0.873 \times 52.6}{0.206}$ (2)

....

....

....

TOTAL 5

---

28 Simplify the following surds, giving your answers in the form $p + q\sqrt{r}$.

a) $(2 + \sqrt{3})^2$

....

.... (2)

b) $\dfrac{12}{\sqrt{3}}$

....

.... (2)

c) $\dfrac{1}{2 + \sqrt{3}}$

....

....

.... (3)

TOTAL 7

ANSWERS ON PAGE 22 ANSWERS ON PAGE 22 ANSWERS ON PAGE 22 ANSWERS ON PAGE 22

29 $a = \sqrt{2}$, $b = \sqrt{8}$, $c = \sqrt{5}$.

Work out the values of these expressions.

a) ab

.... ①

b) $\dfrac{ac}{b}$

.... ②

c) $(a + b)^2$

....

.... ②

TOTAL 5

---

30 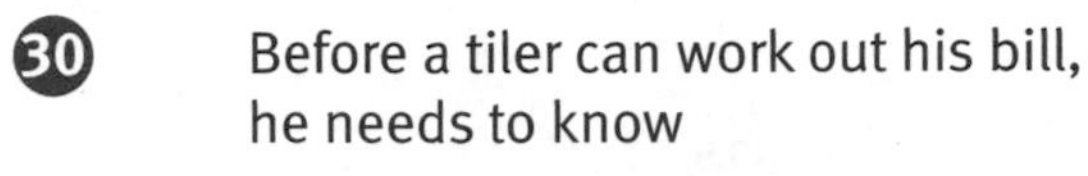

Before a tiler can work out his bill,
he needs to know

- the area to be tiled
- the cost of the tiles per square metre
- how long the job will take.

He makes a profit of 15% on the cost of the tiles.
His labour charge is £20 an hour.
VAT at 17.5% must be added to the total bill.
He deducts 15% for cash.

Work out how much he will charge for laying tiles on $12\,m^2$, with tiles costing £19.63 per m^2.
The job takes 11 hours and the bill is paid in cash.

....

....

....

....

....

.... ⑤

TOTAL 5

ANSWERS ON PAGE 22 ANSWERS ON PAGE 22 ANSWERS ON PAGE 22 ANSWERS ON PAGE 22

QUESTION BANK ANSWERS

1 a) $2 \times 2 \times 2 \times 3$ or $2^3 \times 3$ (2)
b) i) 168 (1)
ii) 8 or $2 \times 2 \times 2$ (1)
c) i) HCF = 12, LCM $12 \times 7 \times 3 = 252$ (4)
ii) HCF is the repeated factor(s). (1)

EXAMINER'S TIP

When dividing to find factors, be systematic, e.g. start with 2. When no more factors of 2, try 3 and so on.

2 $a = 1$, $b = 2$, $c = 2$ (3)

EXAMINER'S TIP

Divide to find the factors as before, then read the indices.

3 a) i) 5 ii) $\frac{1}{9}$ iii) 1 (3)
b) $\frac{1}{2}$ (1)
c) 3^{10} (2)

EXAMINER'S TIP

Remember the negative index is the same as reciprocal.

4 a) $3\sqrt{10} - 2\sqrt{10} = \sqrt{10}$ (2)
b) $1\frac{1}{2}$ (1)
c) 60 (1)

EXAMINER'S TIP

Look for the perfect squares in the surds, e.g., 9, 4.

5 £2800 (1)

EXAMINER'S TIP

Think of the calculation to find the total including the VAT and then do the inverse.

6 a) 1.988×10^9 ($1.988\,1288 \times 10^9$) (2)
b) 1.735×10^{10}($1.734\,576\,497 \times 10^{10}$) (3)

EXAMINER'S TIP

In each part you will lose a mark if you do not round to 4 significant figures.

7 a) $2^4 \times 3^2$ (3)
b) i) 40 or $2^3 \times 5$ (2)
ii) 16 or 2^4 (1)
c) i) $\frac{1}{4}$ (1)
ii) $\frac{1}{125}$ (1)
iii) 8 (1)

EXAMINER'S TIP

A fractional index means a root.

8 a) $4\frac{7}{20}$ (3)
b) $4\frac{2}{5}$ (3)

EXAMINER'S TIP

$\frac{22}{5}$ is acceptable in b) but $\frac{88}{20}$ will lose 1 mark.

9 a) $\frac{7}{12}$ (3)
b) $\frac{51}{44}$ or $1\frac{7}{44}$ (3)

EXAMINER'S TIP

When dividing (or multiplying) change the mixed numbers into top-heavy fractions.

10 a) i) 0.1875 (1)
ii) $0.2\dot{6}$ (2)
b) i) $\frac{3}{32}$ (2)
ii) $\frac{4}{111}$ (2)

EXAMINER'S TIP

Take care when dividing down $\frac{9375}{100\,000}$ not to lose count of the 5s.

11 $\frac{1}{4}$ and $\frac{7}{12}$ (4)

EXAMINER'S TIP

Rather than guess, you could use algebra to find these fractions, $x + y = \frac{10}{12}$, $x - y = \frac{4}{12}$

12 a) $0.4\dot{3}$ (2)
b) $\frac{1}{6} + \frac{4}{15} = \frac{26}{60}$ (3)

13 $101.5 \times \pi \times \left(\frac{222.5}{2\pi}\right)^2 = 399868$ mm^3 or 399.9 cm^3 (3)

EXAMINER'S TIP

The lower bound is needed for each measurement in this case.

14 £374.47 (3)

EXAMINER'S TIP

Remember to divide by 1.175.

15 a) 1.445 m (2)
b) 5 (2)

EXAMINER'S TIP

Don't forget the first two bounces.

16 a) £15 700 ❸
b) 6 more years ❸

EXAMINER'S TIP

This time, don't add in the first year.

17 a) £149 899.75 ❸
b) £149 696.22 ❷

EXAMINER'S TIP

Careful counting is required in b) as you $\times$ 1.005 and $-$ 800.

18 a) £23 090, £11 545, £19 241 ❸
b) 10 : 1 ❷

EXAMINER'S TIP

It is sensible to give these answers to the nearest pound.

19 a) $1^2 - (\sqrt{2})^2 = 1 - 2 = -1$ ❷
b) $\frac{3\sqrt{2}}{\sqrt{3}\sqrt{2}} = \sqrt{3}$ ❷
c) $16 - 24\sqrt{2} + 9 \times 2 = 34 - 24\sqrt{2}$ ❸

EXAMINER'S TIP

Don't forget that $(\sqrt{2})^2$ is 2!

20 3333, 1667, 1250 ❹

EXAMINER'S TIP

It is sensible to round answers up as if you round down, there will not be enough blocks.

21 a) £123.60 ❸
b) £140.74 ❸

EXAMINER'S TIP

In b), notice that the deductions are half the pay after deductions. You need $\frac{7}{33}$ of this.

22 6072 ❸

23 22.5 kg ❸

EXAMINER'S TIP

Remember to multiply by $(0.75)^{10}$.

24 1219 cm^3 ❸

EXAMINER'S TIP

The error is 0.5 mm so maximum length is 20.05 cm.

25 No – could be 463 kg to 465 kg ❺

EXAMINER'S TIP

The maximum possible error is 2.5 kg on each.

26 172.57... 173.58... ❹

EXAMINER'S TIP

For maximum speed use maximum distance divided by minimum time.

27 a) 14 (10 to 15) ❸
b) 200 to 250 ❷

EXAMINER'S TIP

There are fact marks for showing the approximations you make, so write them down.

28 a) $7 + 4\sqrt{3}$ ❷
b) $4\sqrt{3}$ ❷
c) $2 - \sqrt{3}$ ❸

EXAMINER'S TIP

In c), multiply the top and bottom by $2 - \sqrt{3}$.

29 a) 4 ❶
b) $\frac{\sqrt{5}}{2}$ ❷
c) 18 ❸

30 £490.23 ❺

Number

CHAPTER 2

Algebra

To revise this topic more thoroughly, see Chapter 2 in *Letts Revise GCSE Mathematics Study Guide.*

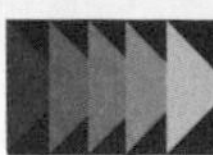

Try this sample GCSE question and then compare your answers with the Grade C and Grade A model answers on the next page.

1 a Find the point of intersection of the straight lines with equations

$$y = x + 5$$
$$x + y = 3$$

....

.... **[3]**

b

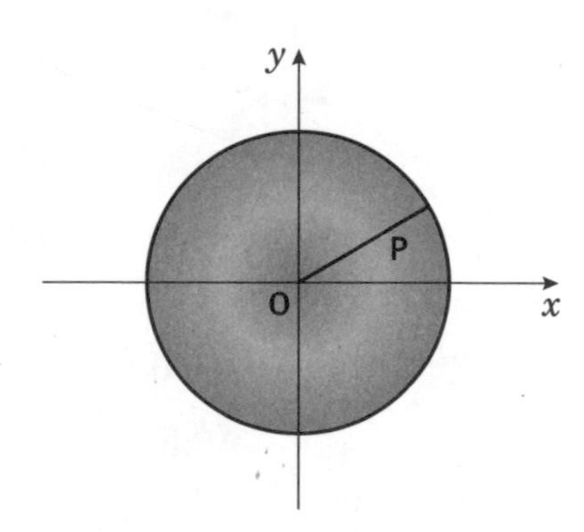

This circle has its centre at the origin and radius 5.

Prove that its equation is

$$x^2 + y^2 = 25.$$

....

.... **[3]**

c Find the points of intersection with the circle of the lines

(i) $y = x + 5$ **(ii)** $x + y = 3$

....

....

....

.... **[9]**

(Total 15 marks)

These two answers are at grades C and A. Compare which one your answer is closest to and think how you could have improved it.

GRADE C ANSWER

Nasreen

The first line is wrong. It should be $x - y = -5$.

a $y = x + 5$ so $x - y = 5$ ✗

$x + y = 3$

$2x + y - y = 8$ ✓

$y = 4,\ x = 1$

Following that error, the rest of the work has been correctly done so there is 1 mark for the method.

b $x^2 + y^2 = (x + y)^2$ ✗

But $x = 1$ and $y = 4$, so $= 25$ – proved.

This is not a proof of the result and it contains a serious (and common) error. $(x + y)^2 = x^2 + 2xy + y^2$

c (i) ?

(ii) ?

1 mark = Grade C answer

Grade booster ···> move a C to a B

Nasreen will be lucky to make a C grade unless other questions are much better than this. Silly mistakes in **a** and **b** have lost her 5 marks. As she did not attempt **c**, perhaps she had not studied the equation of the circle in any depth. High grades can only be achieved if you have covered all the work in the course.

GRADE A ANSWER

Charles

a (i) Substitue $y = x + 5$ in $x + y = 3$ ✓

$x + x + 5 = 3$

$2x = -2$, so $x = -1$ ✓

$y = x + 5 = 4$ ✓

The point is $(-1, 4)$

This is well set out and uses the method of substitution, which is most appropriate as one equation is in the form $y =$.

b This is true by Pythagoras

This is not a proof of the result.

c (i) $x^2 + (x + 5)^2 = 25$ ✓

$2x^2 + 10x + 25 = 25$

$2x^2 + 10x = 0$ ✓

$x + 5 = 0$, so $x = -5$ and $y = 0$ ✓

The start is correct, substituting again for y, and the quadratic is right.

However, 2 marks have been lost by not factorising and finding both roots.

(ii) $x + y = 3$ gives $y = 3 - x$ ✓

$x^2 + (3 - x)^2 = 25$

$2x^2 - 6x + 9 = 25$

$2x^2 - 6x - 16 = 0$ ✓

$x^2 - 3x - 8 = 0$

$x = \frac{3 \pm \sqrt{9 - 4 \times -8}}{2} = 4.70$ or -1.70 ✓

y is $3 - x = -1.70$ or 4.70 ✓

This is a completely correct method with correct solutions to the equation.

However, the last mark is lost as Charles has not made clear that the points are (4.7, −1.7) and (−1.7, 4.7). The values must be paired.

10 marks = Grade A answer

Grade booster ···> move A to A*

Charles has scored quite well on this question and, if other answers are as good, will probably be awarded an A grade. If he wants an A*, however, he will have to avoid errors like losing a root of the quadratic by division. He will also need to understand what is required for a proof. In this case, if P has coordinates (x, y), then a right-angled triangle is formed with sides $x, y, 5$. Applying Pythagoras gives the result. Make sure you state the reasons.

Try this sample GCSE question and then compare your answers with the Grade C and Grade A model answers on the next page.

Solve these equations.

a $3x - 11 = 2(5 - x)$

[3]

b $x^2 - 3x - 40 = 0$

[3]

c $\frac{3}{x-2} + \frac{2}{x-3} = 1$

[6]

(Total 12 marks)

These two answers are at grades C and A. Compare which one your answer is closest to and think how you could have improved it.

GRADE C ANSWER

Bill

a $3x - 11 = 10 - 2x$ ✓

$5x - 11 = 10$ ✓

$5x = 21$

Completely correct. This is a Grade C question. → $x = 4\frac{1}{5}$ ✓

b ~~$(x-3)(x-4)$~~ $8 \times 5 = 40$ ← *The 8 and 5 are correct but these factors give $x^2 + 3x - 40$.*

~~$(x+3)(x+4)$~~

$(x-5)(x+8) = 0$ ✗

The solutions are right for Bill's factors, so earn two marks. → $x = 5$ or $x = -8$ ✓✓

c $\frac{3}{x-2} + \frac{2}{x-3} = 1$? ← *No attempt but this is A* work.*

5 marks = Grade C answer

Grade booster ····> move a C to a B

Continuing like this, Bill will get a good grade C and might make B if he avoids errors like this one in part b. He clearly needs more practice in factorising quadratics.

GRADE A ANSWER

Eileen

a $3x - 11 = 2(5 - x)$

$5x = 21$ ✓✓

Completely correct. → $x = 4\frac{1}{5}$ ✓

b $x^2 - 3x - 40 = 0$

Try 2×20✗, 4×10✗, 5×8✓ ✓

$(x-5)(x+8)$✗

$(x+5)(x-8)$✓ ✓

$x = -5$ or 8 ✓ ← *This is Grade B work, so a Grade A candidate would be expected to get this right.*

c $\frac{3}{x-2} + \frac{2}{x-3} = 1$

Common denominator $(x-2)(x-3)$ ✓

$\frac{3(x-3) + 2(x-2)}{(x-2)(x-3)} = 1$

All correct to this point and well laid out! → $\frac{5x - 13}{(x-2)(x-3)} = 1$ ✓

Therefore $5x - 13 = 1$ ✗ ← *Here Eileen 'forgot' the denominator, as though it was 0 rather than 1 on the right.*

$x = 2\frac{3}{5}$ ✗

8 marks = Grade A answer

Grade booster ····> move A to A*

The last part is difficult at Grade A, so Eileen may still get this grade if other work is up to the standard. To be awarded A* you need to be able to solve equations like this last one. Eileen should have multiplied each side by the denominator and rearranged the quadratic for solution.

QUESTION BANK

1 Simplify these expressions.

a) $a^2 \times a^5$

.... ①

b) $\dfrac{a^2}{a^5}$

.... ①

c) $3ab^2 \times 2a^3b$

....

.... ②

d) $\dfrac{8a^3b^4}{2a^4b^3}$

....

.... ②

e) $(3a^2b^3)^2$

....

.... ②

f) $\left(\dfrac{ab^3}{a^3b}\right)^{\frac{1}{2}}$

....

.... ②

TOTAL 10

2 The frequency, *f*, of a radio signal is inversely proportional to its wavelength, *w*.

a) Write this using the $\propto$ symbol and as a formula.

....

....

....

.... ②

b) When the frequency is 198, the wavelength is 1500. Find:

i) the wavelength for a frequency of 720

....

....

.... ②

ii) the frequency for a wavelength of 3.3.

....

....

.... ②

TOTAL 6

ANSWERS ON PAGE 44 ANSWERS ON PAGE 44 ANSWERS ON PAGE 44 ANSWERS ON PAGE 44

3 Match these proportion relationships to the graphs.

a) $y \propto x$ b) $y \propto x^2$ c) $y \propto \frac{1}{x}$ b) $y \propto \frac{1}{x^2}$

i)

ii)

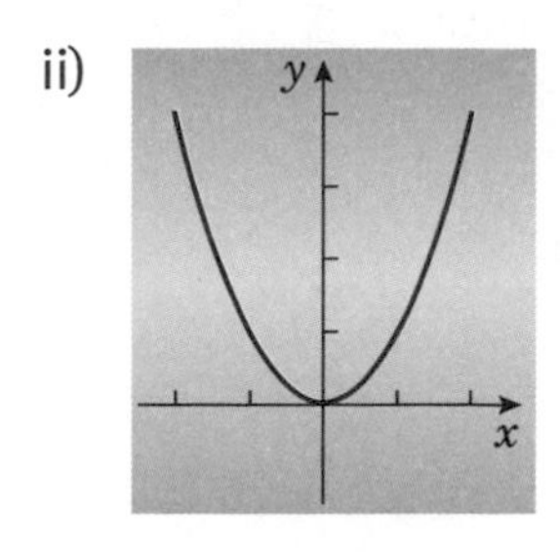

iii)

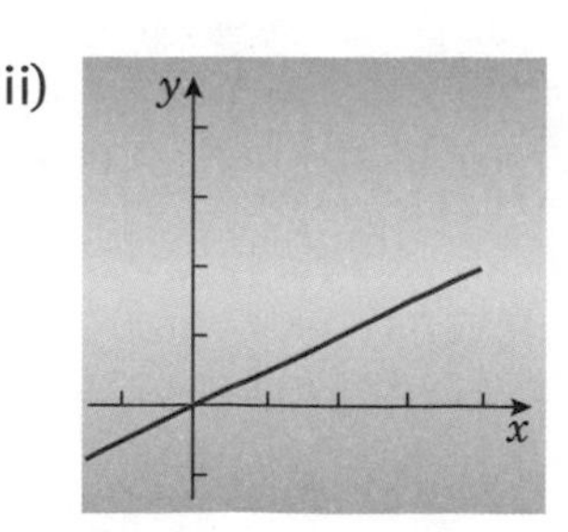

iv)

v)

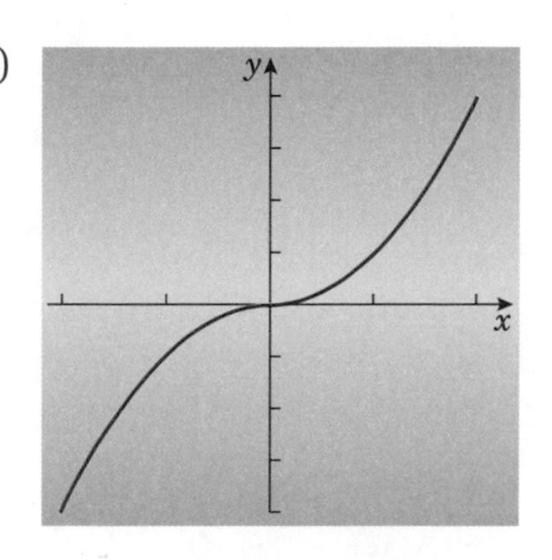

vi)

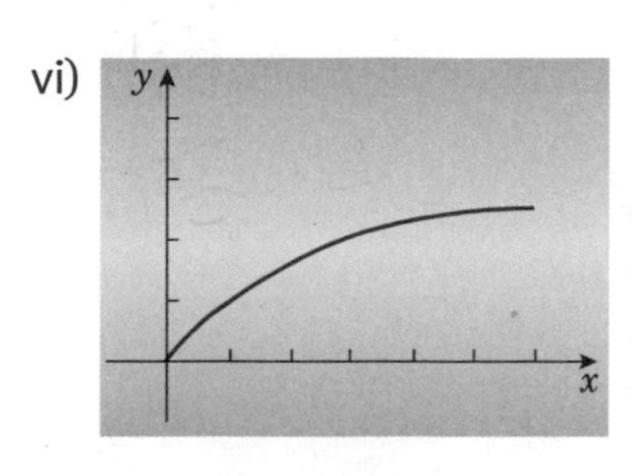

vii)

viii)

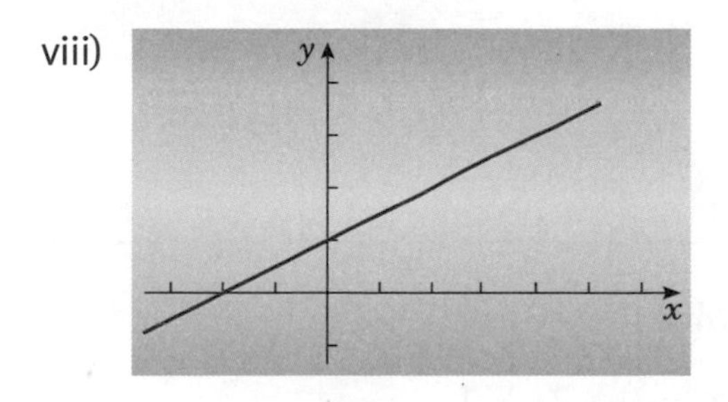

TOTAL 4

4 Look at these expressions.

$$x \quad \frac{1}{x} \quad x^{\frac{1}{2}} \quad x^{-2} \quad x^{-\frac{1}{2}}$$

a) Write the expressions in ascending order when $x > 1$.

....

.... ③

b) Write the expressions in ascending order when $x < 1$.

....

.... ①

TOTAL 4

5 a) Expand the brackets and simplify where possible.

i) $3x(5 - 4x)$

....

.... ①

ii) $(3x + 2)(5 - 4x)$

....

....

.... ②

b) Factorise these expressions.

i) $2x^2 - 4x$

....

.... ②

ii) $x^2 - 7x + 12$

....

.... ②

iii) $6x^2 + 5x - 4$

....

.... ②

TOTAL 9

---

6 Solve these equations.

a) $\frac{1}{2}(x - 3) = 2x + 1$

....

....

.... ③

b) $\frac{x-1}{2} = \frac{x-2}{3}$

....

....

....

.... ③

TOTAL 6

ANSWERS ON PAGE 44 ANSWERS ON PAGE 44 ANSWERS ON PAGE 44 ANSWERS ON PAGE 44

7 A rectangle has one side four times as long as the other side. Its area is $9\,\text{cm}^2$.

Use algebra to find its perimeter.

....

(4)

TOTAL 4

8 Find the points of intersection of the line with the circle.

$2y = x - 4$

$x^2 + y^2 = 81$

....

(5)

TOTAL 5

ANSWERS ON PAGE 44 ANSWERS ON PAGE 44 ANSWERS ON PAGE 44 ANSWERS ON PAGE 44

9 Prove that

a) $\frac{3}{2x-1} - \frac{1}{1-x} = \frac{5x-2}{2x^2-3x+1}$

(4)

TOTAL 4

---

10 $\frac{1}{f} = \frac{1}{u} + \frac{1}{v}$

a) Make f the subject of the formula.

(2)

b) Make v the subject of the formula.

(2)

TOTAL 4

---

11 $P = V(1 + t^2)$

a) Make V the subject of the formula.

(1)

b) Make t the subject of the formula.

(3)

TOTAL 4

---

12 A cylinder has circumference C and height h.

a) Find a formula for its volume V.

....

....

....

....

.... ④

b) Make C the subject of the formula.

....

....

.... ②

TOTAL 6

---

13 $$s = \frac{p}{100(q - p)}$$

Make p the subject of the formula.

....

....

....

....

....

.... ④

TOTAL 4

---

14 Solve these simultaneous equations using algebra.

$$x + 4y = 0$$
$$3x - 2y = -7$$

....

....

....

....

....

....

....

.... ③

TOTAL 3

ANSWERS ON PAGE 44 ANSWERS ON PAGE 44 ANSWERS ON PAGE 44 ANSWERS ON PAGE 44

15 Use a graphical method to solve these equations simultaneously.

$$x - y + 1 = 0$$
$$2x + y - 6 = 0$$

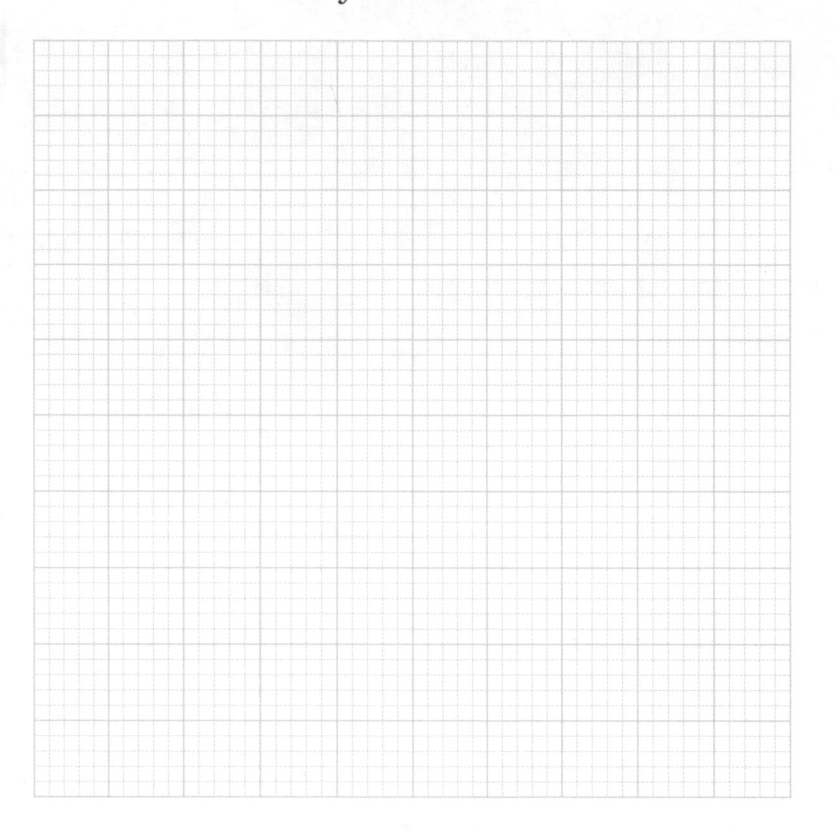

④

TOTAL 4

16 a) Solve these equations simultaneously.

$$x + 2y = 5$$
$$x^2 + y^2 = 36$$

....

....

....

....

....

.... ⑤

b) Give a geometrical interpretation of the result.

....

.... ②

TOTAL 7

ANSWERS ON PAGE 44 ANSWERS ON PAGE 44 ANSWERS ON PAGE 44 ANSWERS ON PAGE 44

17 Use an algebraic method to find the point of intersection of the lines

$$2x + 3y = 7 \quad \text{and} \quad 3x - 2y = 23\tfrac{1}{2}.$$

....

....

....

....

....

....

....

.... ④

TOTAL 4

---

18 Solve these inequalities.

a) $3x - 5 < 10$

....

....

.... ②

b) $6 - 4x \geqslant 2x - 3$

....

....

.... ②

c) $x^2 < 100$

....

....

.... ②

TOTAL 6

ANSWERS ON PAGE 44 ANSWERS ON PAGE 44 ANSWERS ON PAGE 44 ANSWERS ON PAGE 44

Algebra

19 a) Show these inequalities on a graph.

$$y \geqslant 2x - 1 \qquad x + y \leqslant 5$$

④

b) Show by shading the region which represents the solution set of

$$x \geqslant 0, y \geqslant 2x - 1, x + y \leqslant 5.$$

②

c) What is the largest value of x in the region?

.... ①

TOTAL 7

20 a) Factorise

$$x^2 - 2x - 35.$$

....

....

.... ②

b) Hence solve the equation

$$x^2 - 2x - 35 = 0.$$

....

.... ①

c) Hence solve the inequality

$$x^2 - 2x - 35 \leqslant 0.$$

....

....

.... ②

TOTAL 5

ANSWERS ON PAGE 44 ANSWERS ON PAGE 44 ANSWERS ON PAGE 44 ANSWERS ON PAGE 44

21 Use trial and improvement to find the negative root of the equation

$$x^3 - 3x^2 + 5 = 0.$$

Show all the outcomes of your trials and give the answer correct to one decimal place.

....

....

....

....

....

.... (4)

TOTAL 4

---

22 a) The expression

$$2x^2 + 6x - 1$$

may be written as $a(x + b)^2 + c$.
Find the values of a, b and c.

....

....

....

....

.... (3)

b) Hence solve the equation $2x^2 + 6x - 1 = 0$.

....

....

....

.... (3)

TOTAL 6

---

23 Solve these equations.

a) $2x^2 + 5x - 12 = 0$

....

....

....

.... (3)

b) $2x^2 + 5x - 11 = 0$

....

....

....

.... (3)

TOTAL 6

ANSWERS ON PAGE 44 ANSWERS ON PAGE 44 ANSWERS ON PAGE 44 ANSWERS ON PAGE 44

24 Solve the equation

$$\frac{1}{2x+1} - \frac{1}{x+2} = 1$$

....

....

....

....

....

....

....

....

....

....

....

.... ⑥

TOTAL 6

---

25 a) *n* is an integer.
Explain why $2n - 1$ is an odd integer. ①

b) Use algebra to prove

i) the sum of two consecutive integers is odd,

....

....

....

.... ②

ii) the product of two consecutive integers is even

....

....

....

.... ②

TOTAL 5

ANSWERS ON PAGE 44 ANSWERS ON PAGE 44 ANSWERS ON PAGE 44 ANSWERS ON PAGE 44

26 a) Find the nth term of the sequence

$$1 \times 2,\ 2 \times 3,\ 3 \times 4,\ 4 \times 5,\ 5 \times 6,\ \ldots$$

....

....

.... ②

b) Explain why this sequence has the same nth term.

$$(1 + 1),\ (2 + 4),\ (3 + 9),\ (4 + 16),\ (5 + 25), \ldots$$

....

.... ②

TOTAL 4

---

27 a) i) Find the gradient of the straight line shown on the graph.

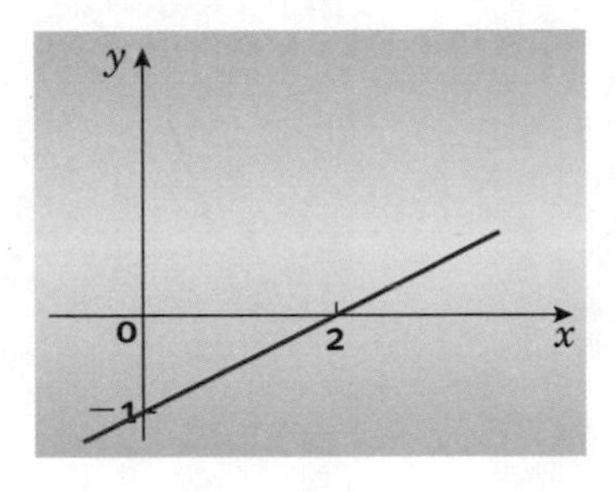

....

....

.... ②

ii) Find the equation of the straight line.

....

....

.... ②

b) i) Find the equation of the line through 0 parallel to the line in part a).

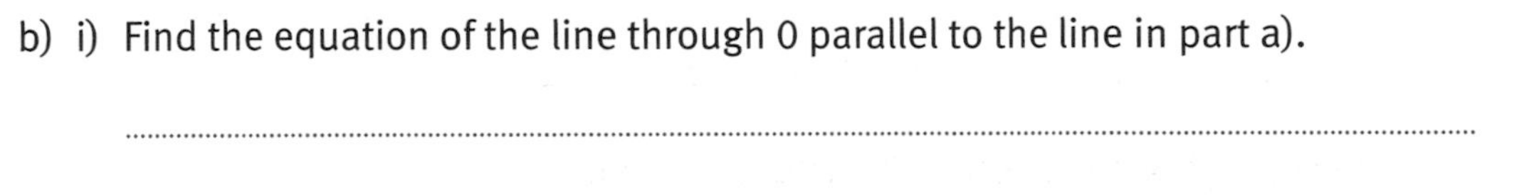

....

....

.... ②

ii) Find the equation of the line through 0 perpendicular to the line in part a).

....

....

....

.... ②

TOTAL 8

ANSWERS ON PAGE 44 ANSWERS ON PAGE 44 ANSWERS ON PAGE 44 ANSWERS ON PAGE 44

28 Water is poured into these containers at the same constant rate.

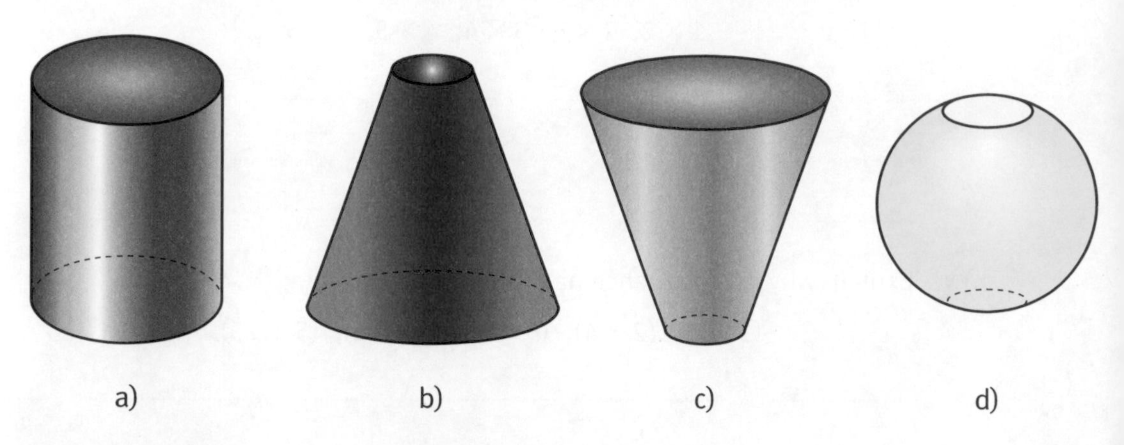

These graphs show the depth of water against time.
Match each container to a graph.

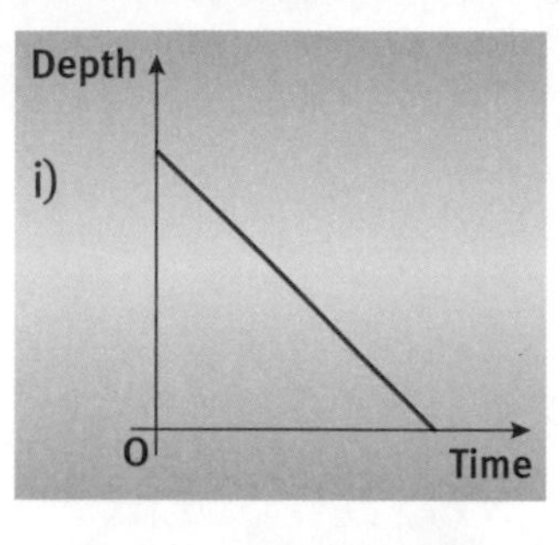

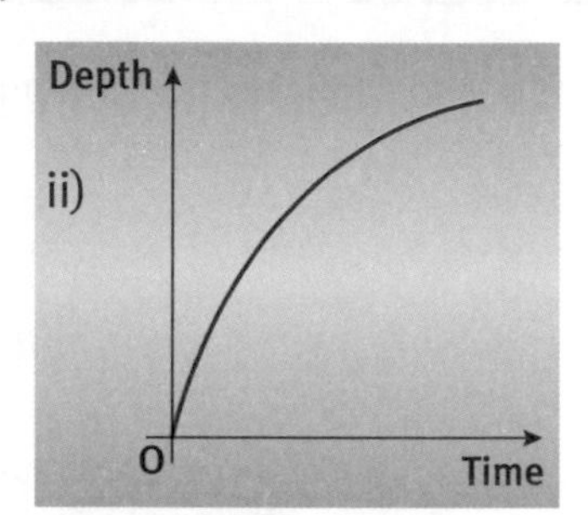

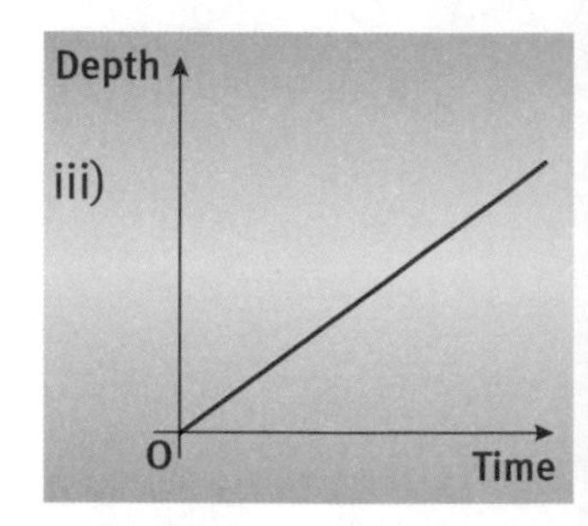

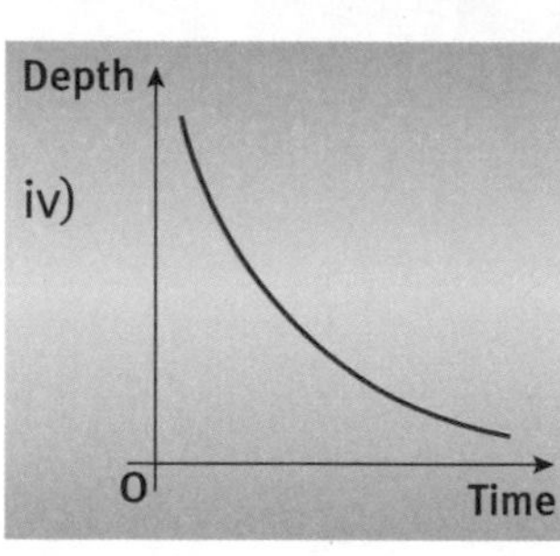

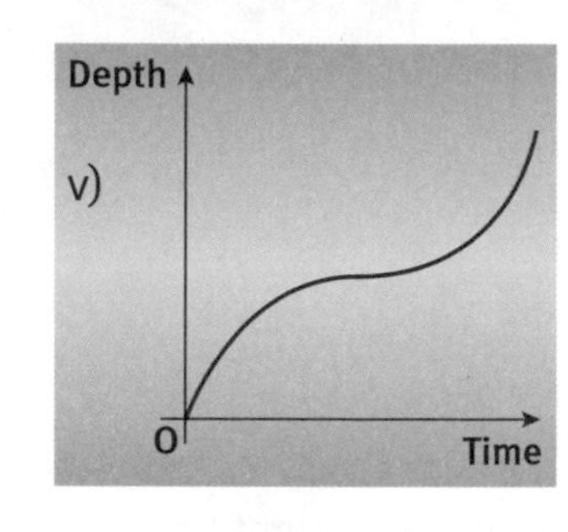

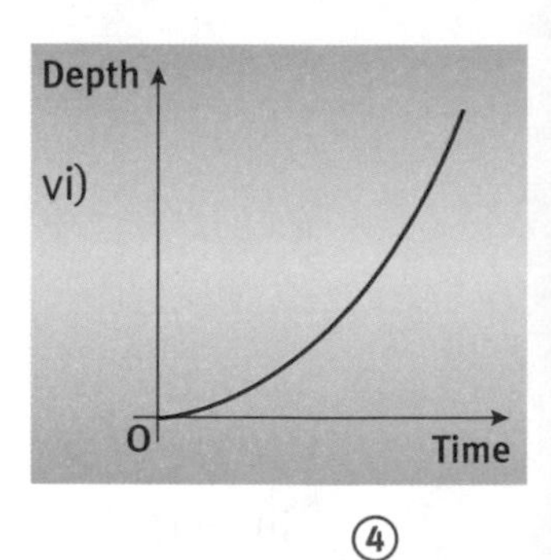

(4)

TOTAL 4

---

29 a) Write down the equations of these loci.

i) Points 2 units from (0, 0).

.... (1)

ii) Points equidistant from (−1, 0) and (3, 0).

.... (1)

b) Find where the two loci intersect.

....

....

....

....

....

.... (4)

TOTAL 6

30 Draw the graph of

$$y = x^3 - 3x + 2,$$

for values of x from -2 to 2.

④

b) From your graph, read off the solutions to

$$x^3 - 3x + 2 = 0.$$

.... ①

c) Use the result from part b) to write $x^3 - 3x + 2$ in factors.

....

....

.... ②

TOTAL 7

ANSWERS ON PAGE 44 ANSWERS ON PAGE 44 ANSWERS ON PAGE 44 ANSWERS ON PAGE 44

Algebra

31 This is the graph of

$$y = x^3 + 4x^2 - 3.$$

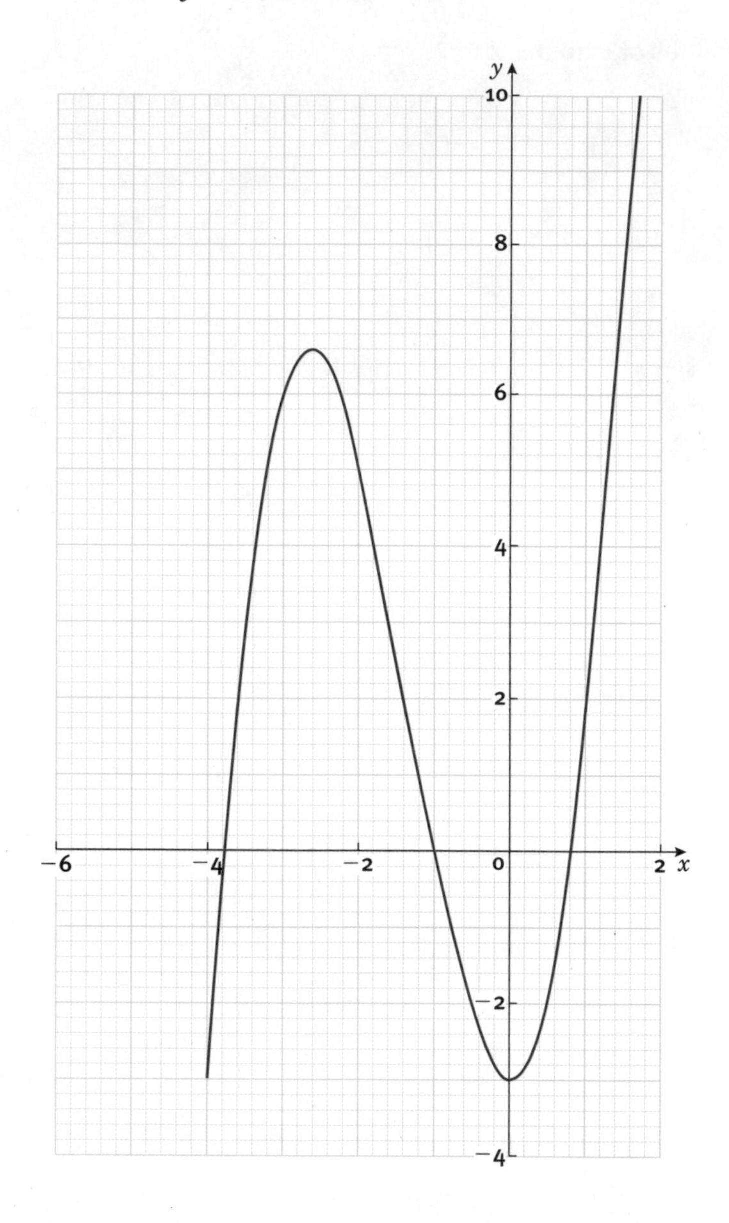

a) Find the roots of the equation $x^3 + 4x^2 - 3 = 0$.

.... ②

b) By drawing suitable straight lines on the graph, solve these equations.

i) $x^3 + 4x^2 - 4 = 0$

....

.... ②

ii) $x^3 + 4x^2 + x - 2 = 0$

....

....

.... ③

TOTAL 7

ANSWERS ON PAGE 44 ANSWERS ON PAGE 44 ANSWERS ON PAGE 44 ANSWERS ON PAGE 44

32 The sketch graph shows $y = \mathrm{f}(x)$.

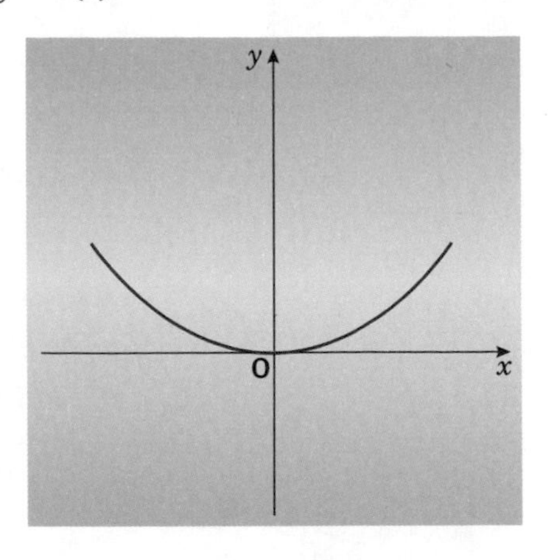

Sketch the graphs of

a) $y = \mathrm{f}(2x)$

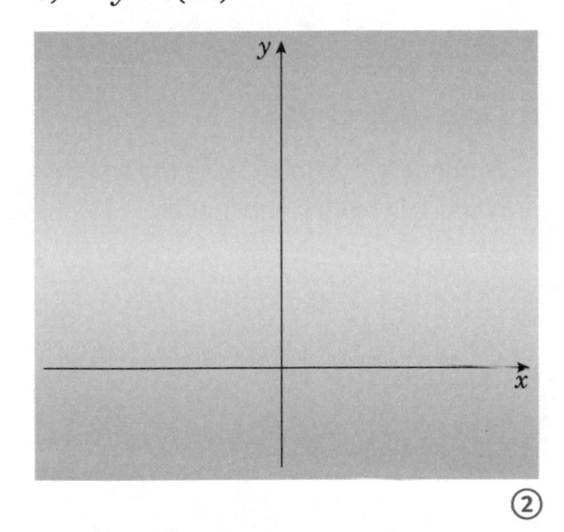

②

b) $y = \mathrm{f}(x) + 1$

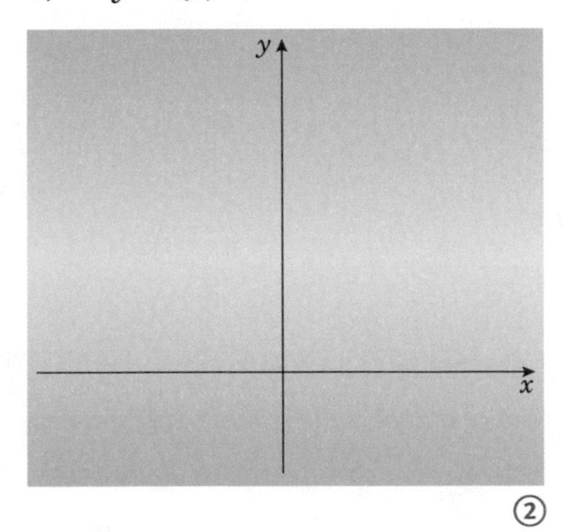

②

c) $y = \mathrm{f}(x + 1)$

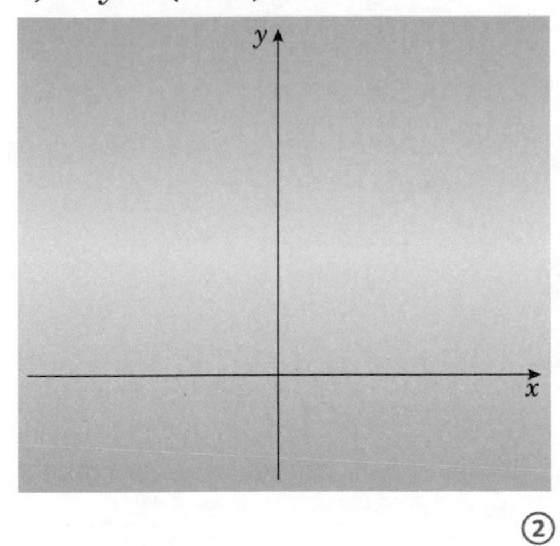

②

d) $y = 2\mathrm{f}(x)$

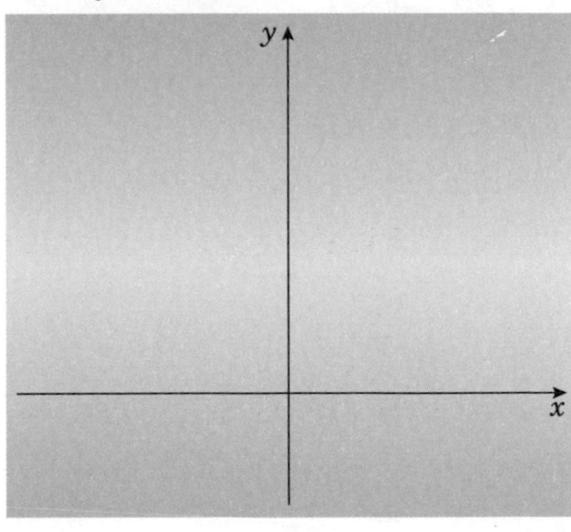

②

TOTAL 8

ANSWERS ON PAGE 44 ANSWERS ON PAGE 44 ANSWERS ON PAGE 44 ANSWERS ON PAGE 44

QUESTION BANK ANSWERS

1 a) a^7 ❶ b) a^{-3} ❶
c) $6a^4b^3$ ❷ d) $4a^{-1}b$ or $\frac{4b}{a}$ ❷
e) $9a^4b^6$ ❷ f) $\frac{b}{a}$ or $a^{-1}b$ ❷

EXAMINER'S TIP

Remember $(b^3)^2$ is b^6.

2 a) $f \propto \frac{1}{w}$; $f = \frac{k}{w}$ ❷
b) i) 412.5 ❷ ii) 90 000 ❷

EXAMINER'S TIP

It may be easier to use $fw = k$ (= 297 000 in this case).

3 a) iii); b) ii); c) i); d) vii) ❹

4 a) $x^{-2}, \frac{1}{x}, x^{-\frac{1}{2}}, x^{\frac{1}{2}}, x$ ❸

b) $x, x^{\frac{1}{2}}, x^{-\frac{1}{2}}, \frac{1}{x}, x^{-2}$ ❶

EXAMINER'S TIP

Notice that b) is the reverse order from a).

5 a) i) $15x - 12x^2$ ❶
ii) $10 + 7x - 12x^2$ ❷
b) i) $2x(x-2)$ ❷
ii) $(x-3)(x-4)$ ❷
iii) $(3x+4)(2x-1)$ ❷

EXAMINER'S TIP

Take care with the signs. Multiply out as a check.

6 a) $x = -1\frac{2}{3}$ ❸ b) $x = -1$ ❸

EXAMINER'S TIP

Signs again!

7 15 cm ❹

EXAMINER'S TIP

*Let the short side be **y** cm, $4y \times y = 9$. Solve the equation.*

8 $(-7.09, -5.54)$, $(8.70, 2.34)$ ❺

EXAMINER'S TIP

Make sure that the coordinates are correctly paired.

9 $\frac{3(1-x)-(2x-1)}{(2x-1)(1-x)} = \frac{-5x+2}{3x-1-2x^2} = \frac{-(5x-2)}{-(2x^2-3x+1)}$ ❹

EXAMINER'S TIP

Don't forget the common denominator and take care with the signs

10 a) $f = \frac{uv}{u+v}$ ❷ b) $v = \frac{uf}{u-f}$ ❷

EXAMINER'S TIP

Combine the two fractions first.

11 a) $V = \frac{P}{1+t^2}$ ❶ b) $t = \sqrt{\frac{P}{V} - 1}$ ❸

EXAMINER'S TIP

Isolate the t^2 before taking the square root. It could be positive or negative.

12 a) $V = \frac{C^2h}{4\pi}$ ❹ b) $C = \sqrt{\frac{4\pi V}{h}}$ ❷

EXAMINER'S TIP

Find r (the radius) first.
$r = \frac{C}{2\pi}$

13 $p = \frac{100sq}{1+100s}$ ❹

EXAMINER'S TIP

Notice that p appears twice. These must be collected.

14 $x = -2, y = \frac{1}{2}$ ❸

15 $x = 1.7, y = 2.7$ (exact $1\frac{2}{3}, 2\frac{2}{3}$) ❹

EXAMINER'S TIP

From a graph you cannot expect greater accuracy.

16 a) $x = -3.98, y = 4.49$; $x = 5.98, y = -0.49$ ❺
b) Points of intersection of line $x + 2y = 5$ with circle radius 6, centre 0 ❷

EXAMINER'S TIP

The easiest way to start is substituting $x = 5 - 2y$

17 $(6\frac{1}{2}, -2)$ ❹

FOR MORE INFORMATION ON THIS TOPIC ... SEE REVISE GCSE MATHS ... CHAPTER 2

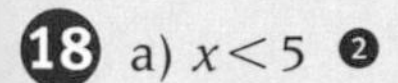

18 a) $x<5$ (2) b) $x\leqslant 1\frac{1}{2}$ (2) c) $-10<x<10$ (2)

EXAMINER'S TIP

Remember the negative square root in c).

19 a) b)

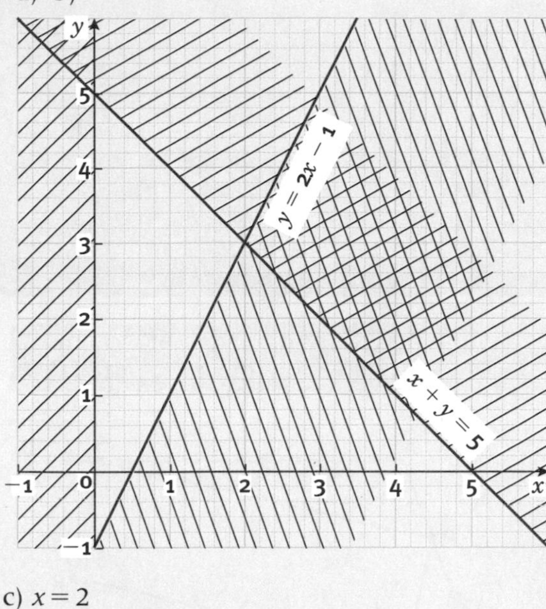

(6)

c) $x = 2$ (1)

EXAMINER'S TIP

It is clearer to shade the unwanted regions.

20 a) $(x-7)(x+5)$ (2) b) $x=7$ or $x=-5$ (1)
c) $-5\leqslant x\leqslant 7$ (2)

EXAMINER'S TIP

For c), one bracket must be positive and one negative.

21 $x=-1.1$ (4)

EXAMINER'S TIP

You should show the root is nearer to −1.1 than −1.2

22 a) $a=2,\ b=1\frac{1}{2},\ c=-5\frac{1}{2}$ (3)
b) $0.16, -3.16$ or $-1\frac{1}{2}\pm\frac{1}{2}\sqrt{11}$ (3)

23 a) $1\frac{1}{2}, -4$ (3) b) $1.41, -3.91$ (3)

EXAMINER'S TIP

Try to factorise first

24 $-0.18, -2.82$ (6)

25 a) $2n$ is even, one less is odd (1)
b) i) $(2n-1)+2n=4n-1$, which is one less than a multiple of 4 (2)
ii) $2n\,(2n-1)=4n^2-2n=2\,(2n^2-n)$, which is $2\times$ an integer (2)

EXAMINER'S TIP

You could choose other forms for consecutive integers, e.g., 2n, 2n + 1.

26 a) $n(n+1)$ (2)
b) $n(n+1)=n^2+n$ (2)

EXAMINER'S TIP

You should work with nth terms to show the result.

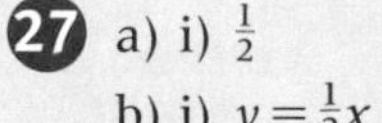

27 a) i) $\frac{1}{2}$ (2) ii) $y=\frac{1}{2}x-1$ (2)
b) i) $y=\frac{1}{2}x$ (2) ii) $y=-2x$ (2)

EXAMINER'S TIP

You can have alternative forms for the equation, e.g. $2y = x$.

28 a) iii); b) vi); c) ii); d) v) (4)

29 a) i) $x^2+y^2=4$ (1) ii) $x=1$ (1)
b) $(1, \sqrt{3}), (1, -\sqrt{3})$ (4)

EXAMINER'S TIP

A sketch will help to spot quick solutions.

30 a)

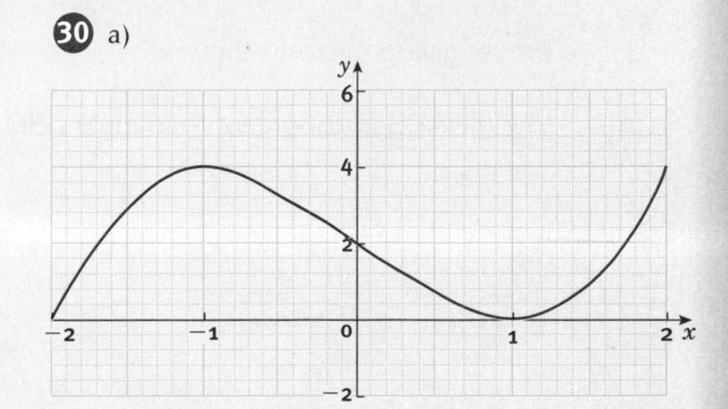

(4)

b) $-2, 1$ (1)
c) $(x+2)(x-1)^2$ (2)

31 Solutions from graph
a) $-3.8, -1, 0.8$ (2)
b) i) $-3.7, -1.2, 0.9$ (read at $y=1$) (2)
ii) $-3.6, -1, 0.6$ (read at $y=x-1$) (3)

EXAMINER'S TIP

Draw the lines $y = 1$ and $y = x - 1$ on the graph.

32 a) Graph half width (2)
b) Translated up 1 (2)
c) Translated 1 left (2)
d) Double height (2)

CHAPTER 3

Shape, space and measures

To revise this topic more thoroughly, see Chapter 3 in *Letts Revise GCSE Mathematics Study Guide.*

Try this sample GCSE question and then compare your answers with the Grade C and Grade A model answers on the next page.

1 The diagram shows a parallelogram ABCD.

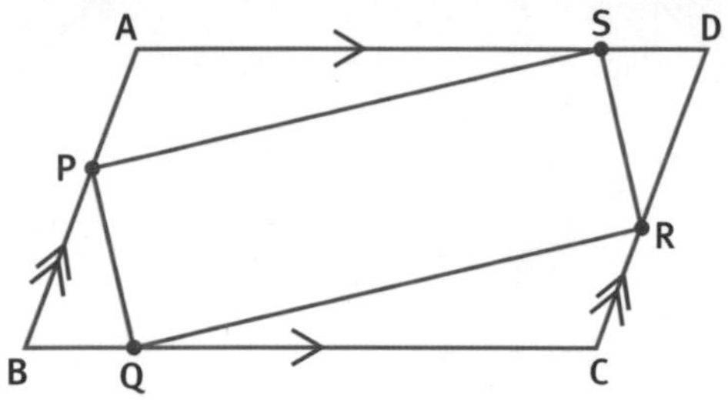

AB = 10 cm and BC = 16 cm.
Points P, Q, R, S are fixed on AB, BC, CD and DA respectively, so that
AP = BQ = CR = DS = 3 cm.

Prove, giving reasons, that

a triangle APS is congruent to triangle CRQ

[3]

b the quadrilateral PQRS is a parallelogram.

[3]

(Total 6 marks)

These two answers are at grades C and A. Compare which one your answer is closest to and think how you could have improved it.

GRADE C ANSWER

Lucy

a Angle PAS + Angle PBQ = 180°

Angle QCR + Angle PBQ = 180°

Therefore Angle PAS = Angle QCR ✓

Angle ASP = angle RQC

Angle APS = angle CRQ ✗

therefore the triangles are congruent. ✗

b Because AS = QC and AP = CR then

PS = QR

also PS is parallel to QR by symmetry

Therefore PQRS is a parallelogram ✗

Although this relationship is derived and earns one mark this is a long way of doing so – the relationship should have been known however.

No reasons are given for these two statements and so no mark. She also assumes that PS is parallel to QR – but this has yet to be established!

Proving that two triangles are congruent depends on proving that, or using the fact that, at least one side in each triangle is equal. What has happened here is that Lucy has used the condition for similarity and therefore gains no marks.

These facts are given but it is not valid to then state that this means PQ = QR. The symmetry argument is incomplete and therefore inconclusive – more information about why symmetry is present is needed. No marks are gained in this part.

1 mark = Grade C answer

Grade booster ···> move a C to a B

Lucy will need to improve her understanding of proof. She will have to appreciate that (i) only facts that are given or proved may be used and (ii) reasons/explanations for statements must be provided – had she done this, for example in part a, she would have gained 1 more mark. She also needs to learn the difference between similarity and congruence.

GRADE A ANSWER

Asif

a AP = CR ✓

angle PAS = angle QCR (interior opposite angles of parallelogram) ✓

AS = QC

Therefore triangle APS is congruent to ✓ triangle CRQ (SAS)

b Because triangle APS is congruent to triangle CQR then angle ASP = angle CQR ✓

Therefore PS is parallel to QR

Therefore PQRS is a prallelogram ✗

(1 pair of opposite sides are parallel.)

The correct reasons are given here so the maximum marks for this part are gained.

Only one valid deduction, the angles being equal, is made from the fact that the triangles are congruent and gains one mark. The fact that PS is parallel to QR is not explained or proved; this is insufficient proof for a parallelogram – the pair of opposite sides must also be equal and this fact needs establishing here.

4 marks = Grade A answer

Grade booster ···> move A to A*

Asif has gained full marks in part b but lost marks in part b through not including a reason why PS is parallel to QR. (The reason is because AS is parallel to QC and angle ASP = angle CQR.)

QUESTION BANK

1 ABC is an isosceles triangle. AC = BC and X and Y are points on AC and BC so that CX = CY.

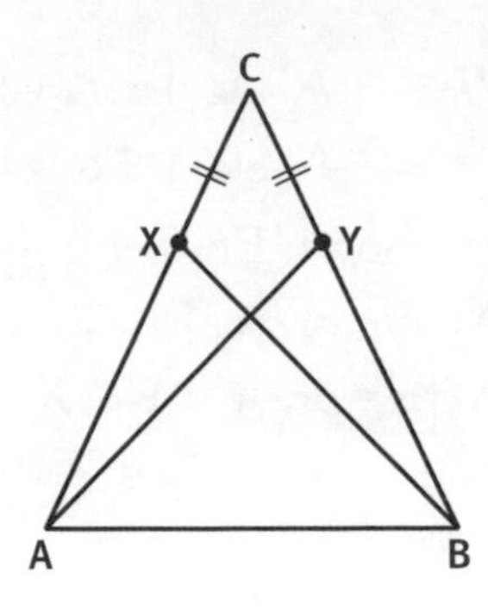

Prove triangle CXB is congruent to triangle CYA.

....

....

....

.... ②

TOTAL 2

2 In this figure DX = XC, DV = ZC, AB is parallel to DC.

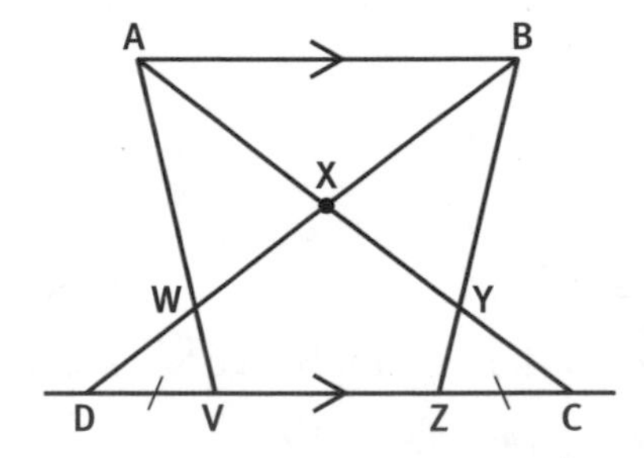

Prove triangle DBZ is congruent to triangle CAV.

....

....

....

....

....

.... ③

TOTAL 3

3 These two triangles are congruent.

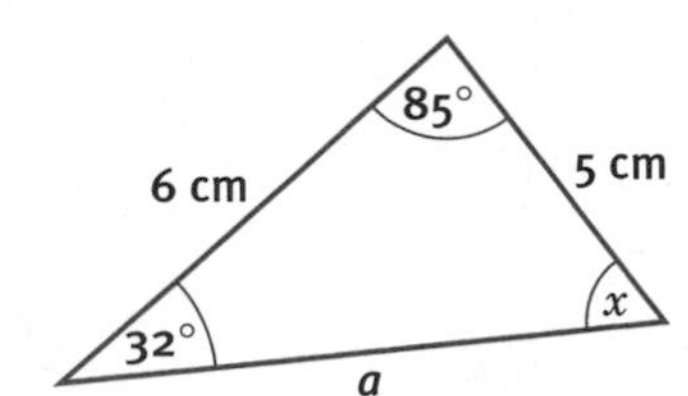

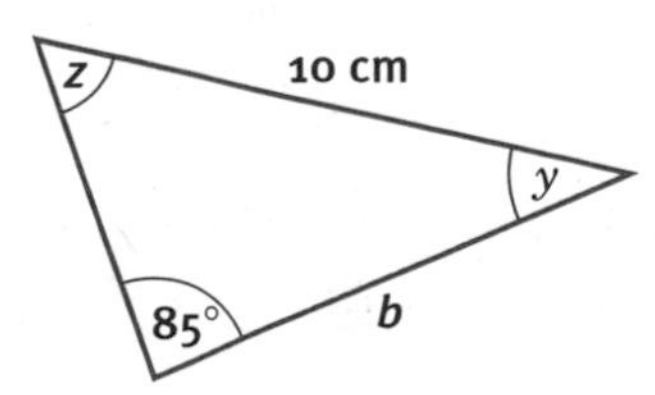

a) Write down the lengths of side a .... cm

side b .... cm ②

b) Write down the sizes of angles x .... °

y .... °

z .... ° ③

TOTAL 5

---

4 ABCD is a rhombus. AB = BC = CD = DA.
AB is parallel to DC, BC is parallel to AD.

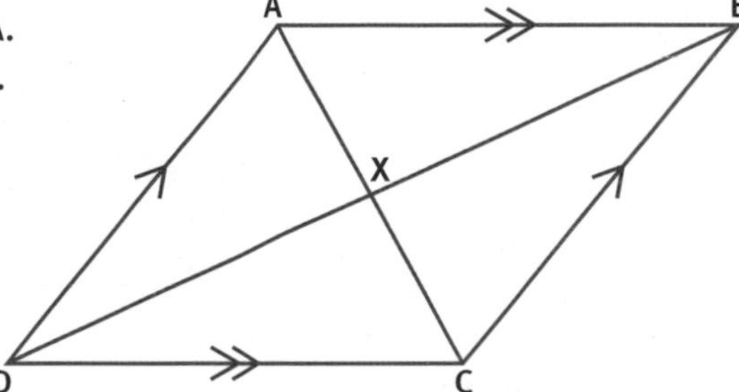

a) Prove that

i) triangles ABX and CDX are congruent

....

....

....

.... ③

ii) triangles ADX and CBX are congruent.

....

....

....

.... ②

b) Prove that X is the midpoint of DB and AC and hence that the diagonals of a rhombus intersect at right angles.

....

....

....

....

.... ③

TOTAL 8

ANSWERS ON PAGE 66 ANSWERS ON PAGE 66 ANSWERS ON PAGE 66 ANSWERS ON PAGE 66

5 In the diagram, AB = AC, angle BAC = 40° and CE = CF.

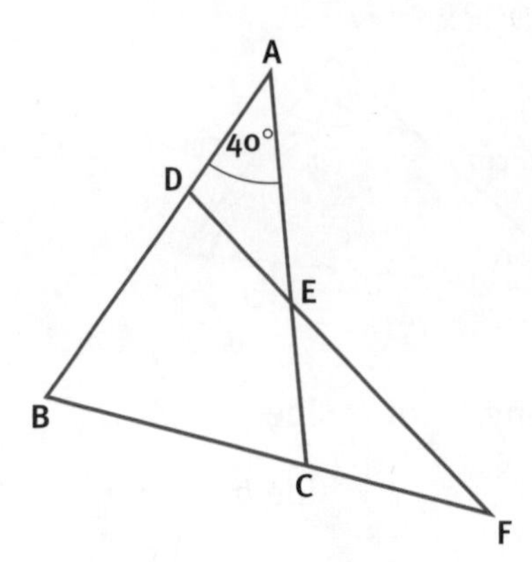

Find angle ADE.

.... (2)

TOTAL 2

---

6 In this diagram, CD is the bisector of angle ACB and AE is parallel to DC.

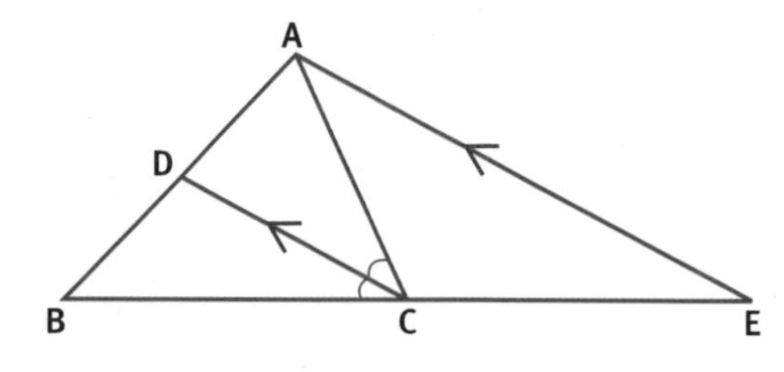

Prove that AC = CE.

....

....

....

....

....

.... (3)

TOTAL 3

---

7 a) Write down $\overrightarrow{AB}$ in terms of $\boldsymbol{a}$ and $\boldsymbol{b}$.

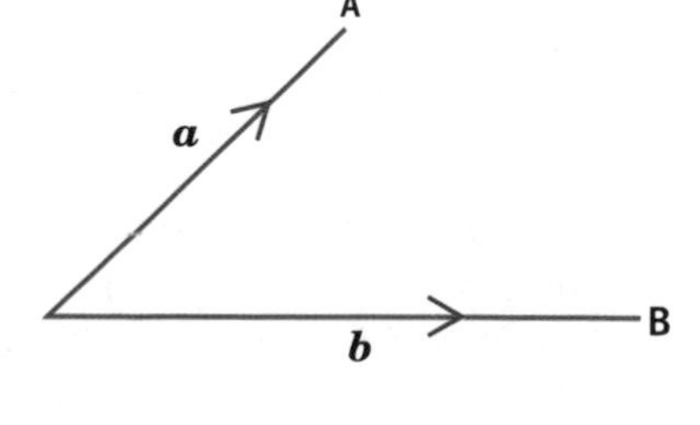

.... (1)

b) P has position vector $\boldsymbol{p} = \boldsymbol{a} + \boldsymbol{b}$.
Prove that OAPB is a parallelogram

....

.... (2)

TOTAL 3

ANSWERS ON PAGE 66 ANSWERS ON PAGE 66 ANSWERS ON PAGE 66 ANSWERS ON PAGE 66

8 ABCD is a quadrilateral. Angle CBA = 70°, angle BAD = 80° and angle ADC = 130°. CA bisects angle DCB.

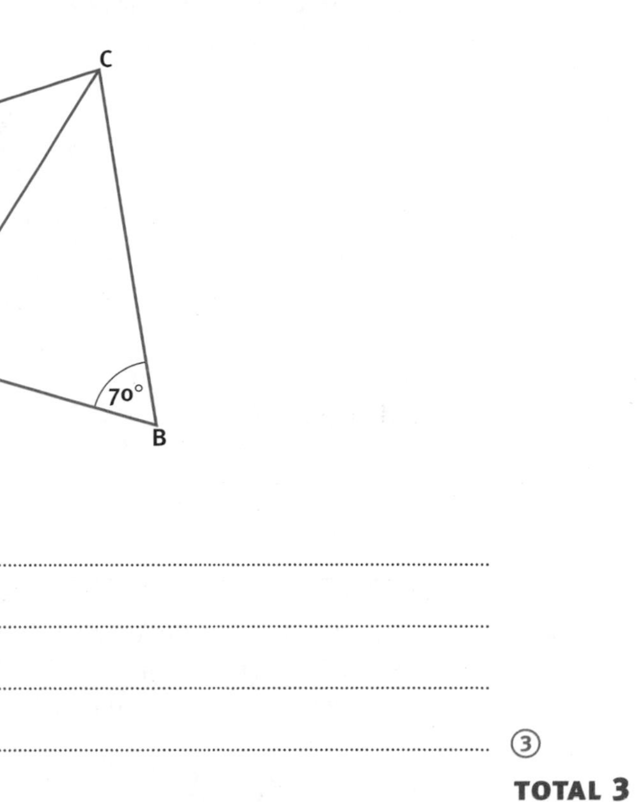

Prove that triangle CAB is isosceles.

....

....

....

.... (3)

TOTAL 3

---

9 Calculate the length of AB in this diagram.

y
B (7, 8)
A (1, 3)
x

....

....

....

.... (2)

TOTAL 2

---

10 a) Sketch the graph of $y = \cos x$ for values of x from 0° to 720°.

.... (2)

b) Find all the values of x in the range 0° to 720° which satisfy $5\cos x = 2$.

Give your answer correct to 0.1°

....

....

.... (3)

TOTAL 5

ANSWERS ON PAGE 66 ANSWERS ON PAGE 66 ANSWERS ON PAGE 66 ANSWERS ON PAGE 66

11 The diagonal of a rectangle is 2 cm greater than the length of the rectangle, l cm. The width of the rectangle is 8 cm.
Calculate the value of l.

....

....

.... (3)

TOTAL 3

12 A, B, C are three points on horizontal ground.
A is 1000 m due west of C and B is 800 m due south of C.
CT is a vertical television mast.

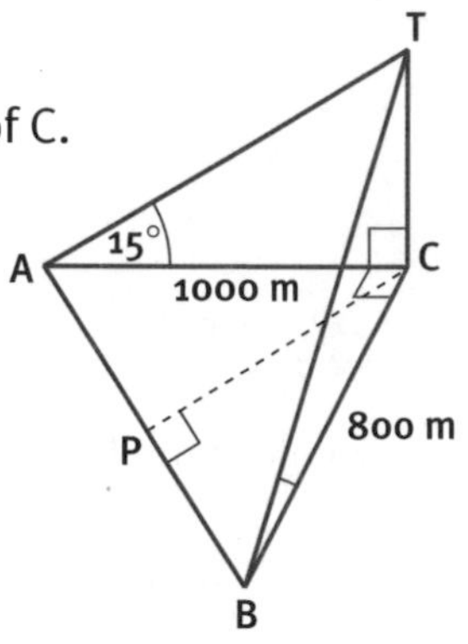

The angle of elevation of T from A is 15°.

a) Find the height of the mast.

....

.... (1)

b) Find the angle of elevation from B.

....

.... (1)

P is the point on AB which is nearest to C.

c) Calculate angle CBA.

....

.... (1)

d) Calculate the distance CP.

....

.... (1)

e) Calculate the angle of elevation of T from P.

....

.... (1)

TOTAL 5

ANSWERS ON PAGE 66 ANSWERS ON PAGE 66 ANSWERS ON PAGE 66 ANSWERS ON PAGE 66

13 The diagram shows two circles inside a square of side 10 cm.
The circle, centre P, has radius r.
The circle, centre Q, has radius 2 cm.

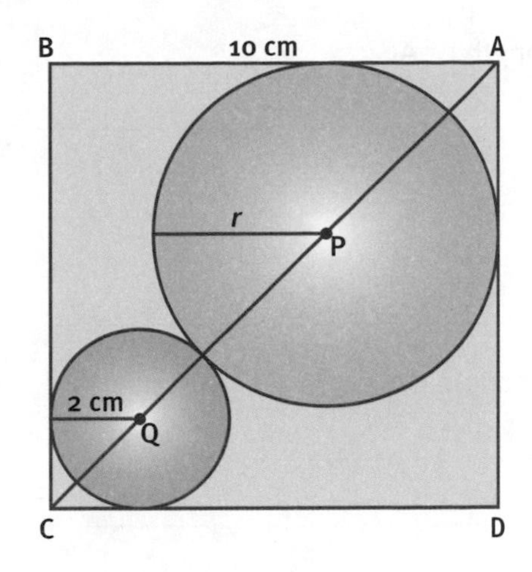

a) Show that $AP = r\sqrt{2}$.

....

....

....

.... ③

b) Write down expressions for AC, PQ, QC.

....

....

.... ③

c) Hence show that $r = \dfrac{8\sqrt{2} - 2}{\sqrt{2} + 1}$

....

....

.... ④

TOTAL 10

14 Points A, B and C lie on a circle centre O.
AB = 20 cm, BC = 15 cm and angle CBA = 30°.

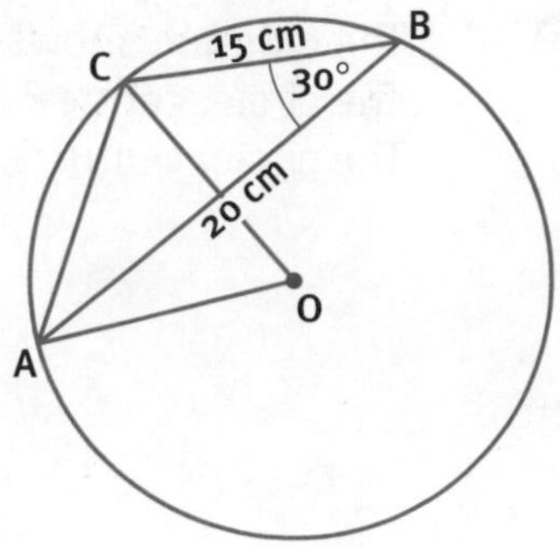

a) Calculate the length of AC.

....

....

....

.... ②

b) Hence calculate angle BAC.

....

....

....

.... ②

c) Prove that triangle AOC is equilateral.

....

....

.... ②

d) Calculate angle OCB.

.... ②

TOTAL 8

---

15 P and Q are two positions, 1000 m apart on horizontal ground, which are used to observe hot air balloons during a race.

The angle of elevation of a balloon from P is 21° and from Q is 25°.

Calculate the height of the balloon.
Give your answer correct to 3 s.f.

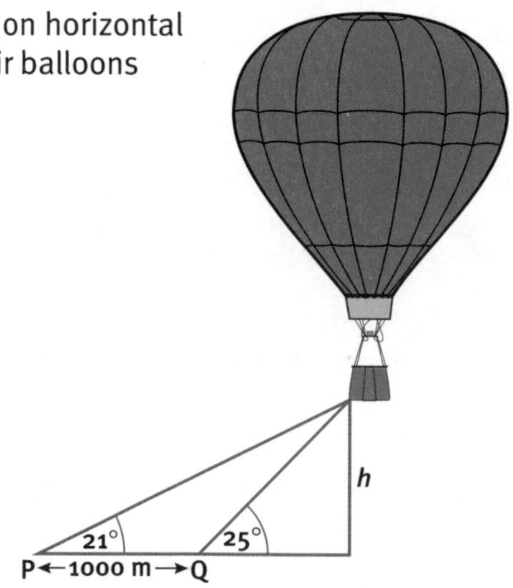

....

....

....

....

....

.... ④

TOTAL 4

16 The sides of the quadrilateral ABCD touch the circle at P, Q, R and S.

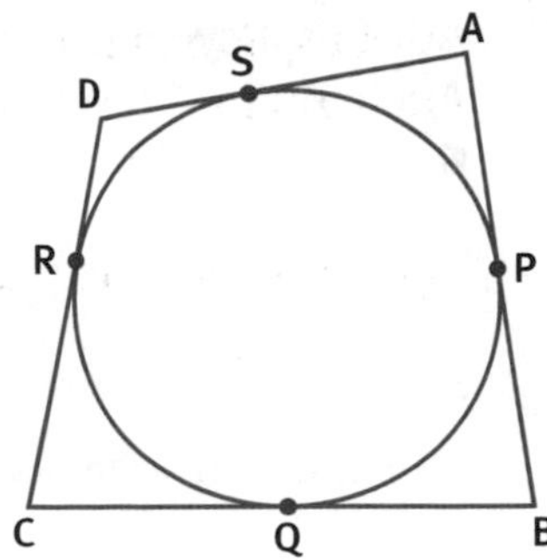

Prove that AB + CD = BC + DA.
Give the reasons for each statement that you make.

....

....

....

....

....

.... (4)

TOTAL 4

17 In the diagram, AT and CT are tangents, B is a point on the circle, angle CAB = 40° and angle BAT = 32°.

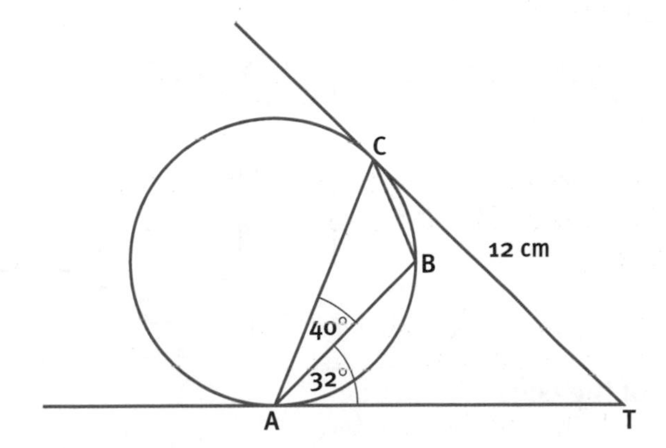

a) i) Calculate the size of angle ACB. Give a reason for your answer.

.... (1)

ii) What is the size of angle BCT?

.... (1)

iii) What is the size of angle CTA?

.... (1)

b) Calculate the length of AC.

....

....

.... (3)

TOTAL 6

ANSWERS ON PAGE 66 ANSWERS ON PAGE 66 ANSWERS ON PAGE 66 ANSWERS ON PAGE 66

18 ABCD is a cyclic quadrilateral. Angle BAC = 40°, Angle BPC = 105° and angle ADB = 30°.

A, B, C, D, P, 40°, 105°, 30°

Calculate the size of angle BCD.

....

....

.... ③

TOTAL 3

19 The volume of a rectangular block of candle wax is 375 cm^3. It has a length of 20 cm and a width of 7.5 cm.

a) Calculate the height of the block.

....

.... ①

The block is melted down and 125 cm^3 of the wax is poured into each of three separate moulds. The first mould is a cube, the second a cylinder of height 12 cm and the third mould is a pyramid with a square base of side x cm and height $2x$ cm.

12 cm, $2x$, x

b) Calculate

i) the length of the edge of the cube

.... ①

ii) the radius of the cylinder

....

....

....

.... ②

iii) the value of x.

....

....

....

.... ②

TOTAL 5

ANSWERS ON PAGE 66 ANSWERS ON PAGE 66 ANSWERS ON PAGE 66 ANSWERS ON PAGE 66

20 A rectangular block of metal has a length of 60 cm, a width of 30 cm and a height of 20 cm.

a) Calculate

i) the volume of the block

.... ①

ii) the total surface area of the block.

....

.... ③

A similar block has length, width and height all 35% greater than the first block.

b) Calculate

i) the volume of the enlarged block

....

.... ②

ii) the total surface area of the enlarged block.

....

.... ②

TOTAL 8

21 In the following expressions, a, b, c, d are lengths.
Indicate which expressions might represent area and which might represent volume.

i) $2(a+b)$

ii) $\frac{1}{3}abc$

iii) $\frac{1}{2}ab\cos\theta$

iv) $4\sqrt{ab+bc+cd}$

v) $\frac{1}{3}(a^2b+ab^2)$

vi) $3a^2$

Area .... Volume ....

....

.... ③

TOTAL 3

22 a) Prove that the area (D) of the triangle is given by

$D = \frac{1}{2} bc \sin A$

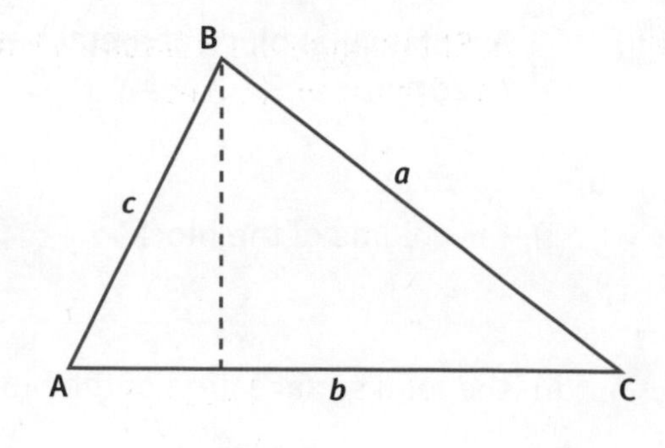

....

.... (2)

b) By writing the formula for the area of the triangle using different sides and angle, prove that

$\frac{a}{\sin A} = \frac{b}{\sin B}$

....

....

.... (3)

TOTAL 5

---

23 A, B and C are the vertices of an equilateral triangle.

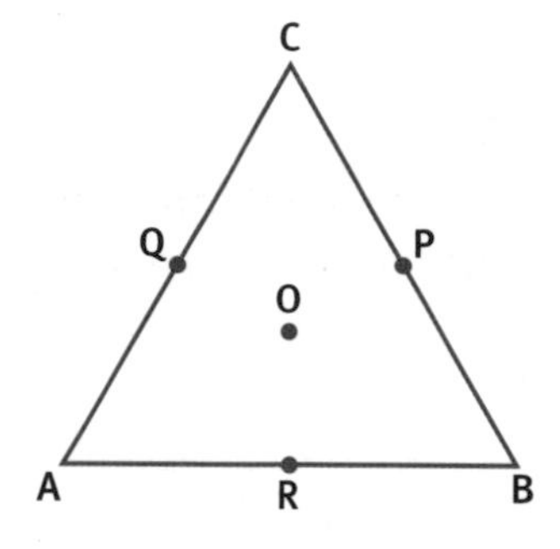

P, Q, R are the midpoints of sides BC, CA, AB respectively.
O is the centre of the triangle.

a) Triangle ABC is rotated anticlockwise through 120° about centre O.
Which points are the images of A, B and C after the rotation?

.... (1)

b) Describe the transformation that will map A onto B, B onto A, and Q onto P.

.... (2)

c) i) Describe the transformation that will map R to Q, C to B, *but not* B to C.

.... (2)

ii) To which point will B be mapped? .... (1)

d) Triangle ABC is enlarged with a scale factor of 2 using A as the centre of enlargement.

What are the images of i) Q .... (1)

ii) R? .... (1)

TOTAL 8

ANSWERS ON PAGE 66 ANSWERS ON PAGE 66 ANSWERS ON PAGE 66 ANSWERS ON PAGE 66

24 'It Sticks' glue is available in two sizes of tubes.
The tubes are similar.

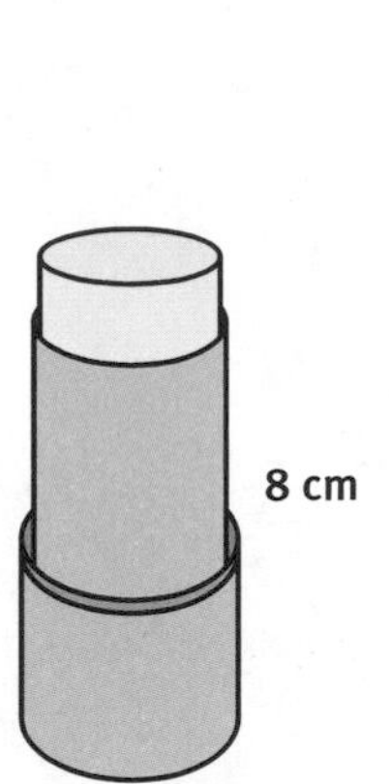

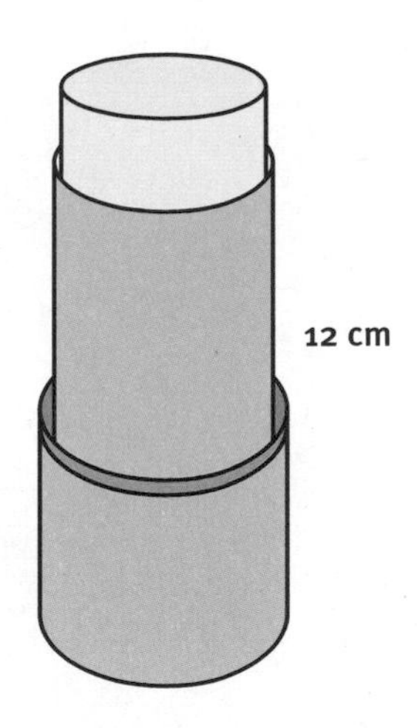

The larger size contains 50 cm^3 of glue.
How much glue is in the smaller size.

....

....

.... ②

TOTAL 2

---

25 A coastguard (C) records the position of a ship at 1100 (S_1) as 10 km on a bearing of 330°. One hour later, he records the position of the same ship (S_2) as 15 km on a bearing of 052°.

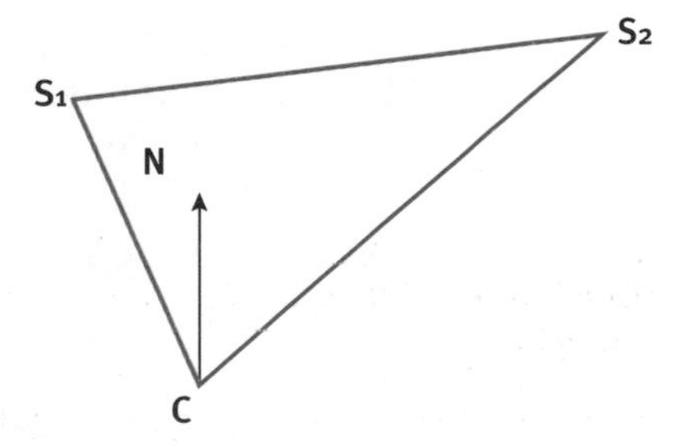

a) How far had the ship travelled?

....

.... ③

b) The ship travelled in a straight line. What was the bearing?

....

.... ③

TOTAL 6

ANSWERS ON PAGE 66 ANSWERS ON PAGE 66 ANSWERS ON PAGE 66 ANSWERS ON PAGE 66

26 In this diagram, $\overrightarrow{OA} = \mathbf{a}$, $\overrightarrow{OC} = 3\mathbf{a}$ and $\overrightarrow{OB} = 2\mathbf{b}$.
E is the midpoint of AB and $BF = \frac{1}{4}BC$.

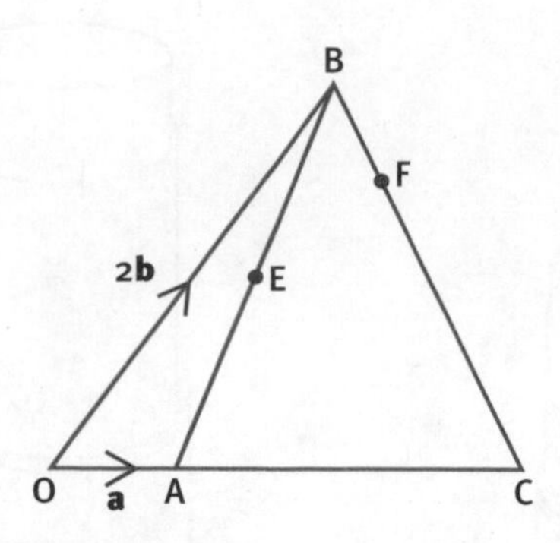

a) Express in terms of **a** and **b**

i) $\overrightarrow{AB}$ .... ①

ii) $\overrightarrow{AE}$ .... ①

iii) $\overrightarrow{OE}$ .... ①

iv) $\overrightarrow{OF}$ ....

.... ①

b) Deduce two facts about O, E, and F.

....

.... ②

TOTAL 6

27 OABC is a parallelogram and $\overrightarrow{OA} = 3\mathbf{a}$ and $\overrightarrow{OC} = 2\mathbf{c}$.
D is the midpoint of AB, E divides the line OA in the ratio 2 : 1 and M is the midpoint of the line OD.

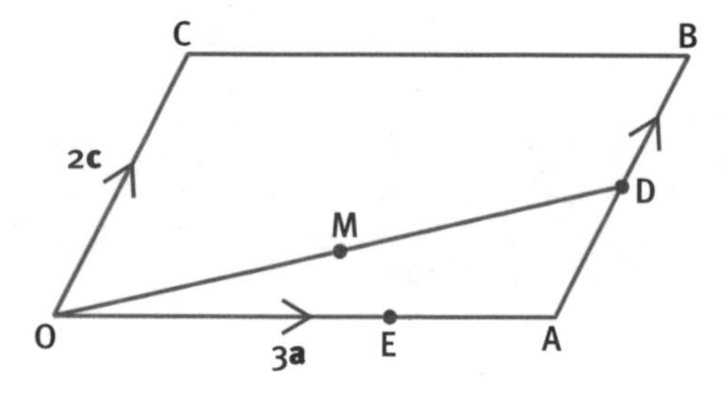

a) Express in terms of **a** and **c**:

i) $\overrightarrow{OE}$ .... ①

ii) $\overrightarrow{OD}$ .... ②

iii) $\overrightarrow{OM}$ .... ①

iv) $\overrightarrow{EC}$ .... ②

b) Prove that M lies on the line EC and that EM : MC = 1 : 3.

....

.... ④

TOTAL 10

ANSWERS ON PAGE 66 ANSWERS ON PAGE 66 ANSWERS ON PAGE 66 ANSWERS ON PAGE 66

28 Here is a sketch of a triangle.

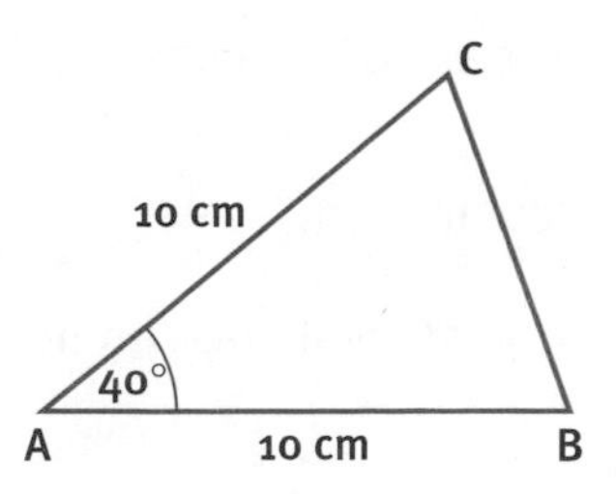

Anna was asked to make an accurate drawing of the triangle.
She can only measure lengths accurately to ± 1 mm and angles to $\pm 1°$.
What is the shortest possible length of the line BC?

....

....

....

....

....

.... (3)

TOTAL 3

---

29 A policeman timed a car travelling along a 100 m section of road. The time taken was 6 seconds.

The length of the road was accurate to the nearest 10 cm, and the time was measured accurate to the nearest second.

What was the greatest speed the car could have been travelling at?

....

....

.... (3)

TOTAL 3

---

30 A town has a population of 74 000, and covers an area of 6.4 km^2.
Calculate the population density of the town. Give your answer correct to 3 s.f.

....

.... (2)

TOTAL 2

ANSWERS ON PAGE 66 ANSWERS ON PAGE 66 ANSWERS ON PAGE 66 ANSWERS ON PAGE 66

31 Amy has bought a new house. She is going to make a lawn in the back garden.

Amy buys a box of 4 kg of grass seed.

The instructions on the box say 100 g of seed will cover between 1 square metre and 1.25 square metres.

Amy decides to make a circular lawn.

What is the radius of the largest lawn she can make? Give your answer correct to 2 s.f.

....

....

....

.... (3)

TOTAL 3

---

32 A large circular sign is fixed into the ground at the entrance to a park.

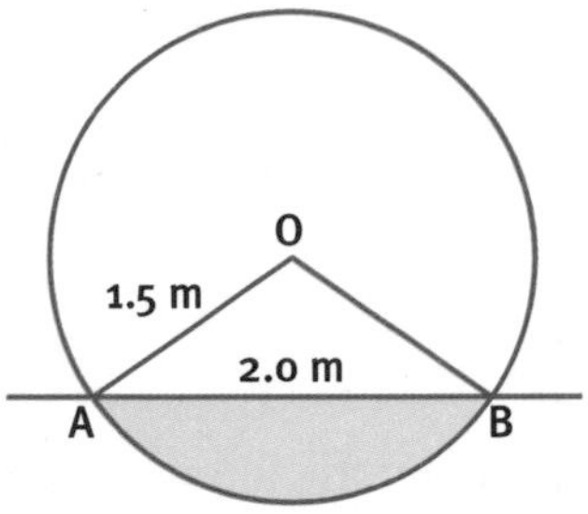

O
1.5 m
2.0 m
A
B

The shaded segment in the diagram shows the part of the sign that is below the surface of the ground.

The radius of the circle is 1.5 m. The chord AB is 2 m.

Calculate the area of the circle which is above the ground. Give your answer correct to 3 s.f.

....

....

....

....

....

.... (5)

TOTAL 5

ANSWERS ON PAGE 66 ANSWERS ON PAGE 66 ANSWERS ON PAGE 66 ANSWERS ON PAGE 66

33 The shot put area on an athletics ground is a sector of a circle. It is marked out in 5-metre intervals.

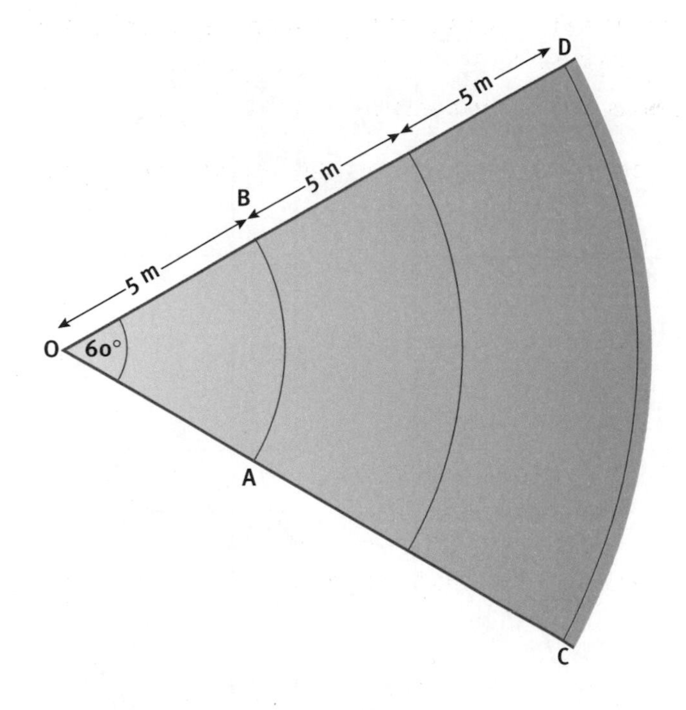

a) Calculate the length of the '5-metre' arc, AB.

....

.... ②

b) Calculate the area of the sector OCD.

....

.... ②

c) It is decided that the area of sector OCD is to be 100 m². What size will angle DOC have to be?

....

....

.... ②

TOTAL 6

ANSWERS ON PAGE 66 ANSWERS ON PAGE 66 ANSWERS ON PAGE 66 ANSWERS ON PAGE 66

34 These are the graphs of six trigonometrical functions.

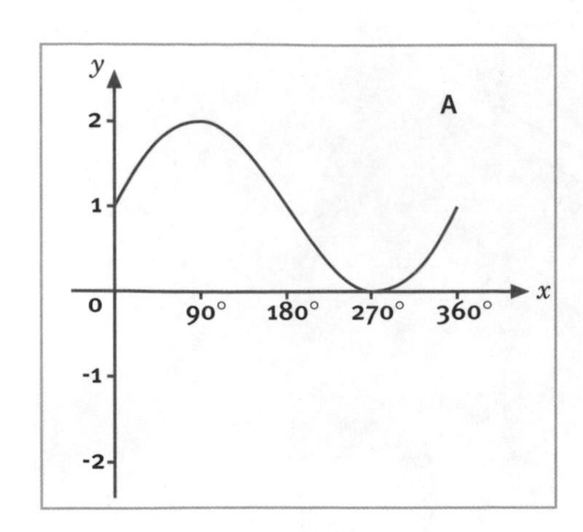

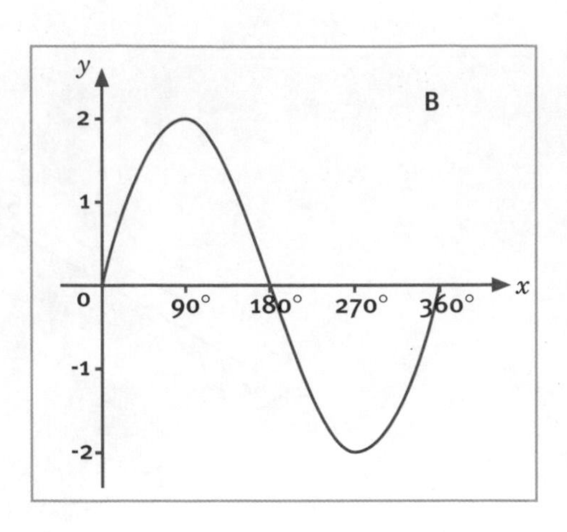

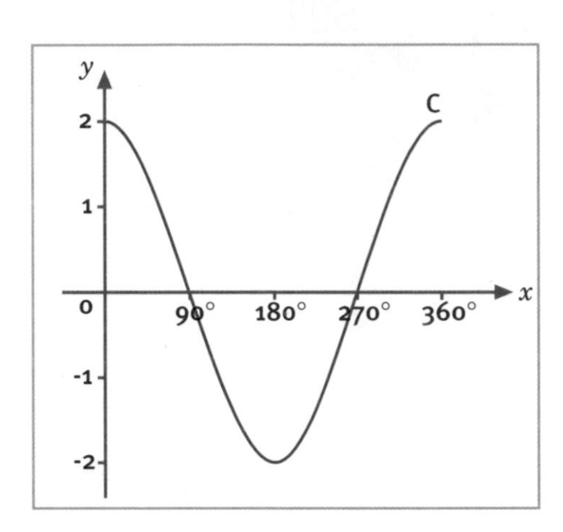

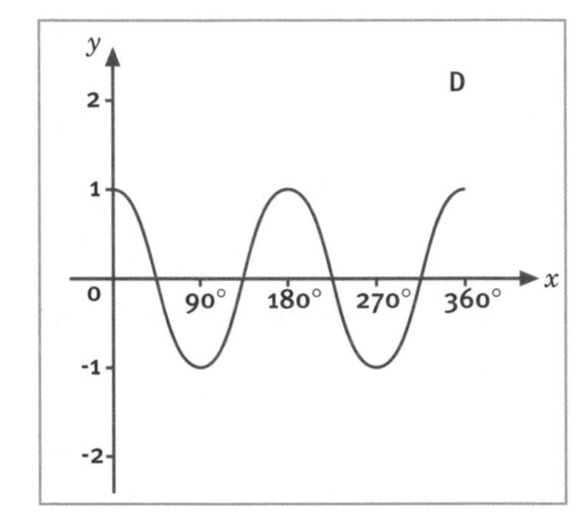

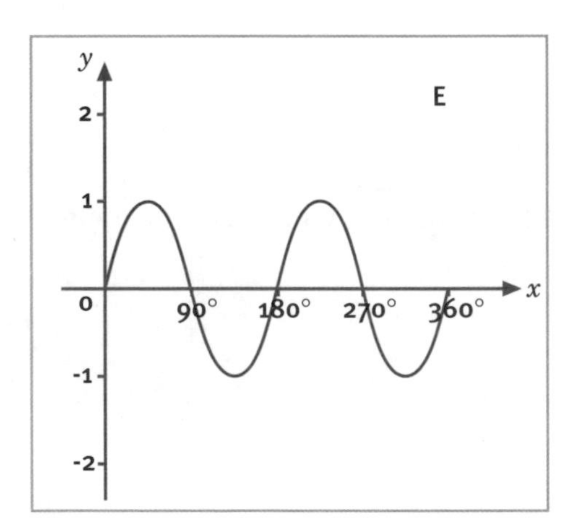

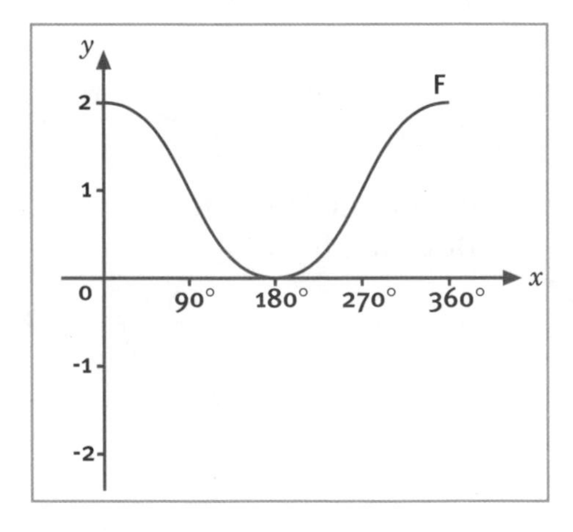

Write the corresponding graph letter (A–F) alongside the trigonometrical functions below.

y = 2sinx .... y = 2cosx .... y = sin2x ....

y = cos2x .... y = 1 + sinx .... y = 1 + cosx .... (4)

TOTAL 4

Two radar stations, A and B, are 150 km apart, with B due south of A.

Station A detects a ship S on a bearing of 146° and finds that the ship is within 100 km of B and nearer to B than A.

Using a scale of 1 cm to 20 km complete a drawing to show the position of the ship S. ③

Do not rub out your construction lines.

TOTAL 3

QUESTION BANK ANSWERS

1
CX = CY; CA = CB (given)
$\angle$C is common
$\therefore \triangle CXB \equiv \triangle CYA$ (SAS) ❷

2
DZ = DV + VZ
CV = CZ + VZ
$\therefore$ DZ = CV
$\triangle$DXC is isosceles (DX = XC, given)
$\therefore \angle BDZ = \angle ACV$
$\therefore \angle XAB = \angle XCD$ (alternate)
and $\angle ABX = \angle XDC$ (alternate)
$\therefore \triangle$AXB is isosceles
$\therefore$ AX = XB
$\therefore$ DB = AC
So $\triangle DBZ \equiv \triangle CAV$ (SAS) ❸

3 a) $a = 10$ cm, $b = 6$ cm, ❷
b) $x = 63°$, $y = 32°$, $z = 63°$ ❸

EXAMINER'S TIP

One triangle is a rotation of the other.

4 a) i) $\angle BAX = \angle XCD$ (alternate)
$\angle ABX = \angle CDX$ (alternate)
AB = DC (given)
$\therefore \triangle ABX \equiv \triangle CDX$ (ASA) ❸
ii) similar proof for $\triangle$s ADX and CBX ❷
b) DX = XB ($\triangle$s AXB and CXD congruent)
AX = XC ($\triangle$s ABX and CDX congruent)
$\therefore$ X is midpoint of DB and AC.
DC = CB (ABCD is a rhombus)
$\therefore \angle XDC = \angle XBC$ ($\triangle$BDC isosceles)
XC is common
$\therefore \triangle DXC \equiv \triangle BXC$
$\therefore \angle DXC = \angle BXC$.
Since DB is a straight line
$\angle DXC = \angle BXC = 90°$ ❸

5
105° ❷

EXAMINER'S TIP

Calculate angle ACB which equals angle CEF plus angle CFE.

6
$\angle DCB = \angle DCA$ (CD bisects $\angle$ACB)
$\angle DCA = \angle CAE$ (alternate)
$\angle DCB = \angle AEC$ (corresponding)
$\therefore \angle AEC = \angle CAE$
$\therefore \triangle$ACE is isosceles and AC = CE ❸

7 a) $\boldsymbol{b} - \boldsymbol{a}$ b) $\overrightarrow{AB} = \boldsymbol{b}$, so AP is equal and parallel to OB. ❸

EXAMINER'S TIP

You could use BP instead. A parallelogram is defined by one pair of opposite sides equal and parallel.

8
$\angle DCB = 80°$ (sum of angles of quadrilateral is 360°)
$\therefore \angle ACB = 40°$ (AC bisects $\angle$DCB)
$\therefore \angle CAB = 180° - (40° + 70°) = 70°$
$\therefore \triangle$CAB is isosceles. ❸

9
7.81 ❷

EXAMINER'S TIP

It helps to treat AB as the hypotenuse of a right-angled triangle.

10 a)

b) 66.4°, 293.6°, 426.4°, 653.6° ❹

EXAMINER'S TIP

Use your sketch in part (a) to help answer part (b).

11
$(l + 2)^2 = l^2 + 8^2$
$l = 15$ cm ❸

EXAMINER'S TIP

Use Pythagoras' theorem to write down the equation $(l + 2)^2 = l^2 + 8^2$.

12 a) 268 m b) 18.5° c) 51.3°
d) 624 m e) 23.2° ❺

EXAMINER'S TIP

It may help to draw the triangles separately.

13 a) Draw radius from P to meet tangent AB at X at right angles.
$\triangle$AXP is isosceles, so AX = XP = r
$AP^2 = r^2 + r^2 = 2r^2$; $AP = r\sqrt{2}$ ❸
b) $AC = 10\sqrt{2}$, $PQ = r + 2$, $QC = 2\sqrt{2}$ ❸
c) AC = AP + PQ + QC
$10\sqrt{2} = r\sqrt{2} + r + 2 + 2\sqrt{2}$
$10\sqrt{2} = (r + 2)(\sqrt{2} + 1)$
$$r = \frac{8\sqrt{2} - 2}{\sqrt{2} + 1}$$ ❹

14 a) 10.3 cm b) 47° ❹
c) $\angle AOC = 60°$ (angle at centre = 2 × angle at circumference) and AO = OC (radii)
$\therefore \triangle$AOC is equilateral ❷
d) 43° ❷

15 2170 m to 3 s.f. (4)

EXAMINER'S TIP

Label the point on the ground, X, therefore $h = QX \tan 25°$ and $h = (1000 + QX) \tan 21°$.

16 AS = AP, BP = BQ, CQ = CR, DR = DS
(tangents from same point are equal)
AB + CD = AP + PB + CR + RD
= AS + BQ + CQ + DS
= BC + DA (4)

17 a) i) 32° (angle in alternate segment) (1)
ii) 40° iii) 36° (2)
b) 7.4 cm (3)

18 95° (3)

EXAMINER'S TIP

Find the angles BDC, BCA using 'angles in the same segment'. Angle BPC is an exterior angle to triangle PDC.

19 a) 2.5 cm (1)
b) i) 5 cm (1) ii) 1.82 cm (2) iii) 5.72 cm (2)

20 a) i) 36 000 cm^3 (1) ii) 7200 cm^2 (3)
b) i) 88 574 cm^3 (2) ii) 13 122 cm^2 (2)

EXAMINER'S TIP

Remember ratio of areas = (ratio of lengths)^2 etc.

21 Area iii), vi) Volume ii), v) (3)

22 a) $D = \frac{1}{2} \times AC \times height = \frac{1}{2}b \times c\sin A$ (2)
b) $D = \frac{1}{2}bc\sin A = \frac{1}{2}ac\sin B$
$b\sin A = a\sin B$ (3)

EXAMINER'S TIP

There is a hint for starting part (a) with the height marked on the triangle.

23 a) B, C and A (1)
b) reflection in the line COR (2)
c) i) rotation of 120° clockwise about O (2)
ii) A (1)
d) i) C ii) B (2)

EXAMINER'S TIP

In part c) it might be easier to think of the transformation that maps R onto Q and see what effect that has on point B.

24 14.8 cm^3 (2)

25 a) 16.8 km (1) b) 088.9° (1)

EXAMINER'S TIP

These are straightforward application of the cosine rule and the sine rule. Keep all the figures in your calculator until the end.

26 a) i) $-\mathbf{a} + 2\mathbf{b}$ ii) $\frac{1}{2}(-\mathbf{a} + 2\mathbf{b})$
iii) $\frac{1}{2}(\mathbf{a} + 2\mathbf{b})$ iv) $\frac{3}{4}(\mathbf{a} + 2\mathbf{b})$ (4)
b) O, E and F are on the same straight line
$OE = \frac{2}{3}OF$ (2)

EXAMINER'S TIP

Take care with signs – be consistent with direction.

27 a) i) $2\mathbf{a}$ (1) ii) $3\mathbf{a} + \mathbf{c}$ (2) iii) $\frac{1}{2}(3\mathbf{a} + \mathbf{c})$ (1)
iv) $2(\mathbf{c} - \mathbf{a})$ (2)
b) $\overrightarrow{EM} = \overrightarrow{EO} + \overrightarrow{OM}$
$= -2\mathbf{a} + \frac{1}{2}(3\mathbf{a} + \mathbf{c})$
$= \frac{1}{2}(\mathbf{c} - \mathbf{a})$
$\overrightarrow{MC} = \overrightarrow{MO} + \overrightarrow{OC}$
$= -\frac{1}{2}(3\mathbf{a} + \mathbf{c}) + 2\mathbf{c}$
$= \frac{3}{2}(\mathbf{c} - \mathbf{a})$
Hence E, M and C lie on the same straight line and $EM : MC = \frac{1}{2} : \frac{3}{2}$
$= 1 : 3$ (4)

EXAMINER'S TIP

Take care with signs – be consistent with direction.

28 6.6 cm (3)

29 19.1 m/s (3)

30 11 600 people per km^2 (2)

31 4.0 m (3)

32 6.54 m^2 (5)

EXAMINER'S TIP

Calculate area of sector AOB and area of triangle AOB.

33 a) 5.24 m b) 118 m^2 c) 51° (6)

34 B, C, E, D, A, F (4)

EXAMINER'S TIP

If in doubt, put some simple values in to check.

35

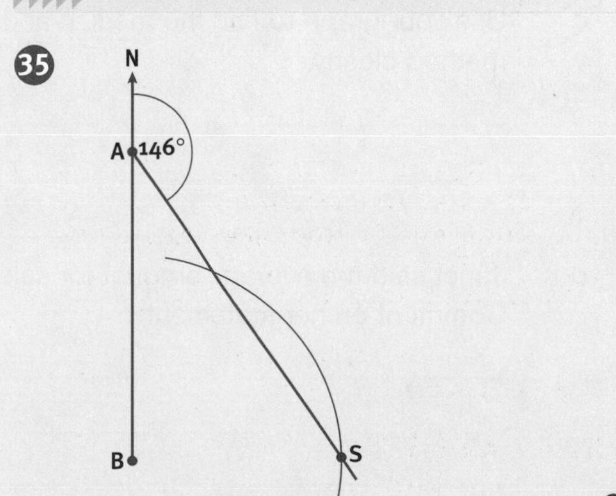

(3)

CHAPTER 4

Handling data

To revise this topic more thoroughly, see Chapter 4 in *Letts Revise GCSE Mathematics Study Guide.*

Try this sample GCSE question and then compare your answers with the Grade C and Grade A model answers on the next page.

People often buy Christmas cards in charity shops. The table shows the money received by a charity shop for two months before Christmas. Note that the shop is only open six days a week.

Money received (£M)	Frequency	Money received (£M)	Frequency
$0 < M \leqslant 50$	30	$200 < M \leqslant 250$	1
$50 < M \leqslant 100$	8	$250 < M \leqslant 300$	1
$100 < M \leqslant 150$	3	$300 < M \leqslant 350$	2
$150 < M \leqslant 200$	3		

a Calculate an estimate of the mean amount received each day.

....

....

.... **[4]**

b Complete the cumulative frequency table and draw the cumulative frequency graph of the distribution.

Money received, (£M)	$\leqslant 0$	$\leqslant 50$	$\leqslant 100$	$\leqslant 150$	$\leqslant 200$	$\leqslant 250$	$\leqslant 300$	$\leqslant 350$
Cumulative frequency	0							

[4]

c Use your graph to find the median and quartiles of the distribution. Show your method clearly.

....

....

.... **[3]**

d Janet said the average amount for sales of cards is about £70.
Comment on her statement.

....

....

.... **[1]**

(Total 12 marks)

These two answers are at grades C and A. Compare which one your answer is closest to and think how you could have improved it.

GRADE C ANSWER

Usha

a (50 × 30 + 100 × 8 + 150 × 3 + 200 × 3 + 250 × 1 + 300 × 1 + 325 × 2) ÷ 61 ✓✗
= £74.59 ✗

Lose one mark for not using the mid-point of each group.

Lose one mark for dividng by 61 (the number of days in 2 months) rather than by 48.

One mark for the correct values.

One mark for a correct division.

b the missing cumulative frequency values are: 30, 38, 41, 44, 45, 46, 48 ✓

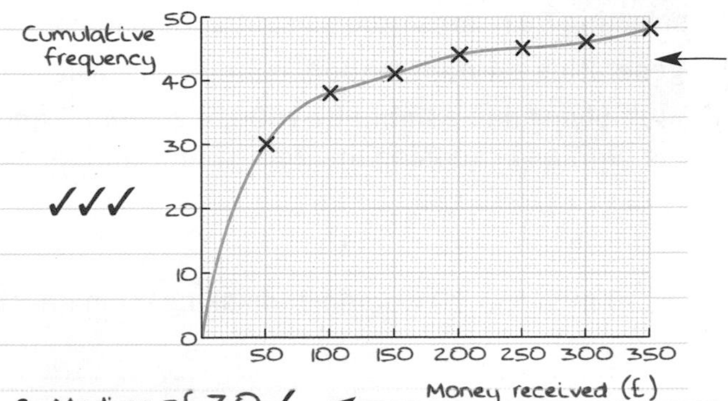

One mark for correctly drawing the axes and labelling the scales.

One mark for plotting the values correctly at the right-hand end of each group.

One mark for drawing a correct curve.

c Median = £30 ✓
lower quartile = £15 ✗
upper quartile = £45 ✗

One mark – generous because there is no indication of how the median was found.

No marks – mis-read scale? or no understanding of quartiles?

d the statement is true because it is close to the mean ✓

*Although the mean has been calculated incorrectly the answer here is justified in relation to the answer to part **a**.*

7 marks = Grade C answer

Grade booster ····> move a C to a B

Usha needs to revise her understanding of how to calculate an estimate of the mean for grouped data. She also needs to be sure she knows how to find the quartiles and to practise reading values on scales.

GRADE A ANSWER

David

a (25 × 30 + 75 × 8 + 125 × 3 + 175 × 3 + 225 × 1 + 275 × 1 + 325 × 2) ÷ 48 ✓✓✓
= £70.83 ✓

***Two marks for correctly using** the mid-points to work out an estimate of the total and 1 mark for dividing by 48*

Two marks for a correct answer

b the missing cumulative frequency values are: 30, 38, 41, 44, 45, 46, 48 ✓

One mark for the correct values.

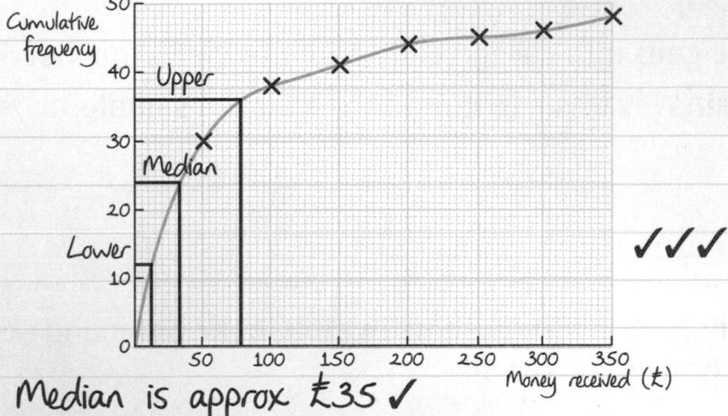

One mark for correctly drawing the axes and labelling the scales,

One mark for plotting the values correctly at the right-hand end of each group,

One mark for drawing a correct curve.

Two marks for indicating how the median was found – drawing a line across at a cum freq of 24

similarly one mark here

and here.

Because there is no reason given, no mark is awarded.

(the statement is only true for the mean. If the median was used the statement would have said '... £35'. So the answer should, for example, be: 'true if the mean is considered'.

c Median is approx £35 ✓
lower quartile is approx £10 ✓
upper quartile is approx £85 ✓
d true

11 marks = Grade A answer

Grade booster ····> move A to A*

David needs to appreciate that reasons, explanations and/or justifications must be given where asked for – and this includes justifying comments as in this question.

QUESTION BANK

1 For each of the following situations describe the type of sampling being used.

a) From a list of names and addresses choosing the 5th, 15th, 25th, 35th name and so on

.... (1)

b) When selecting 90 names from a list of 1800 names:
First generating a 4-digit random number on a calculator, e.g. 0.5247;
then multiplying this number by 1800, e.g. 944.46 and using the integer part e.g. 944 and choosing the 944th name and repeating this 90 times.

.... (1)

c) Stopping 30 men and 30 women and asking them questions.

.... (1)

TOTAL 3

---

2 A group of 800 people was asked for their views on a plan for a new bus station for a city. 450 people said they approved the plan.

a) 190 000 people live in the city.
Estimate the number of people in the city who would support the plan.

.... (2)

b) If this is to be a reliable estimate what assumption must be made about the 800 people who were asked?

.... (1)

TOTAL 3

---

3 Eddie, Bob and Amy are students at college. As part of a project they need to select a random sample of 50 students from the 1350 students at the college.

a) Eddie goes into college early and interviews the first 50 students he sees.
Explain why this might produce a biased sample.

....

.... (2)

b) Bob lists all 1350 students in alphabetic order and selects every 15th on the list until he has chosen 50.

i) Explain why his sample is not random.

.... (2)

ii) State if his sample is likely to be biased.

.... (1)

 ANSWERS ON PAGE 94 ANSWERS ON PAGE 94 ANSWERS ON PAGE 94 ANSWERS ON PAGE 94

c) Amy uses her calculator to generate 3-digit random numbers, such as 0.056 or 0.334.

Assuming that the number generated is always between 0 and 1, explain how Amy could use these numbers to select a random sample.

....

.... ②

TOTAL 7

---

4 Harry decides to investigate the weights of some chocolate bars to test the manufacturer's claim that all the bars weigh 70 g.

Harry buys 100 bars at the same time and from the same shop.

Comment on Harry's method of sampling.

.... ②

TOTAL 2

---

5 The manager of a supermarket wanted to estimate the number of people who visited his store in a town. He telephoned 100 people in the town one evening and asked 'Have you used the supermarket in the last week?'

13 people said they had, so the manager concluded that 13% of the town's population used his supermarket.

Give three criticisms of this method of estimation.

....

....

....

.... ③

TOTAL 3

---

6 The table shows the numbers of boys and the numbers of girls in Year 7 and Year 8 of a school

	Year 7	Year 8
Boys	100	50
Girls	90	60

The headteacher wants to find out what pupils think about wearing school uniform.

A stratified sample of 50 pupils is to be taken from Year 7 and Year 8.

Calculate the number of boys and the number of girls to be sampled from Year 7.

....

....

.... ②

TOTAL 2

ANSWERS ON PAGE 94 ANSWERS ON PAGE 94 ANSWERS ON PAGE 94 ANSWERS ON PAGE 94

Handling data

7 Emily wants to do a survey as part of her statistics coursework. She has produced 30 questionnaires which she wants to give to pupils in each year.

The number of pupils in each year is given in the table below.

Year 7	Year 8	Year 9	Year 10	Year 11
140	140	100	100	120

Emily chooses to use a stratified sample.

How many students from each year should she sample?

....

....

....

....

....

.... (4)

TOTAL 4

---

8 Janet is the manager of a charity shop. She records the daily takings for the shop in four-week blocks.

Table 1 below shows the takings for 2 four-week blocks.

Table 1 Shop takings in pounds

Week	1	2	3	4	1	2	3	4
Day								
Monday	136	140	146	145	145	139	149	165
Tuesday	153	159	168	175	171	194	186	132
Wednesday	146	139	135	150	139	158	164	170
Thursday	162	175	192	204	187	143	190	187
Friday	185	194	238	252	221	240	199	286
Saturday	220	232	275	315	265	270	334	392

a) Why might Janet consider using a 6-point moving average to present the data?

....

.... (1)

ANSWERS ON PAGE 94 ANSWERS ON PAGE 94 ANSWERS ON PAGE 94 ANSWERS ON PAGE 94

The weekly totals are given below in Table 2, together with the first two calculations of the 4-point moving average.

b) Complete the column showing the 4-point moving averages, giving your answers correct to the nearest pound.

Table 2 Total and average takings, in pounds

Week	Total	Moving average
1	1002	
2	1039	
		1109
3	1154	
		1141
4	1241	
		____
1	1128	
		____
2	1144	
		____
3	1222	
4	1332	

③

The graph below shows the weekly totals.

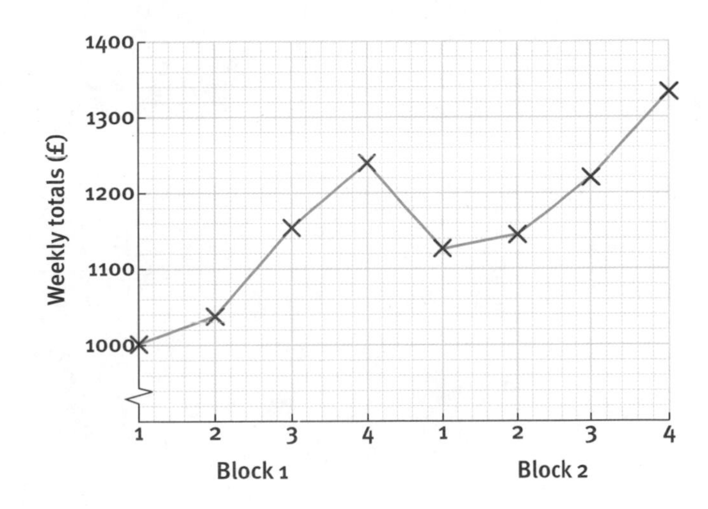

c) Plot the 4-point moving averages on the graph. ②

d) Comment on the general trend and the 4-weekly variation.

....

....

....

.... ③

TOTAL 9

ANSWERS ON PAGE 94 ANSWERS ON PAGE 94 ANSWERS ON PAGE 94 ANSWERS ON PAGE 94

9 The table below shows the numbers of pairs of hiking boots sold by a sports shop during the period 1998–2000.

	Spring	Summer	Autumn	Winter
1998	43	17	15	15
1999	47	19	18	18
2000	57	26	22	13

a) Plot these figures on a graph, using a scale of 1 cm to each season (i.e. quarter) on the horizontal axis and 2 cm to 10 pairs of boots on the vertical axis.

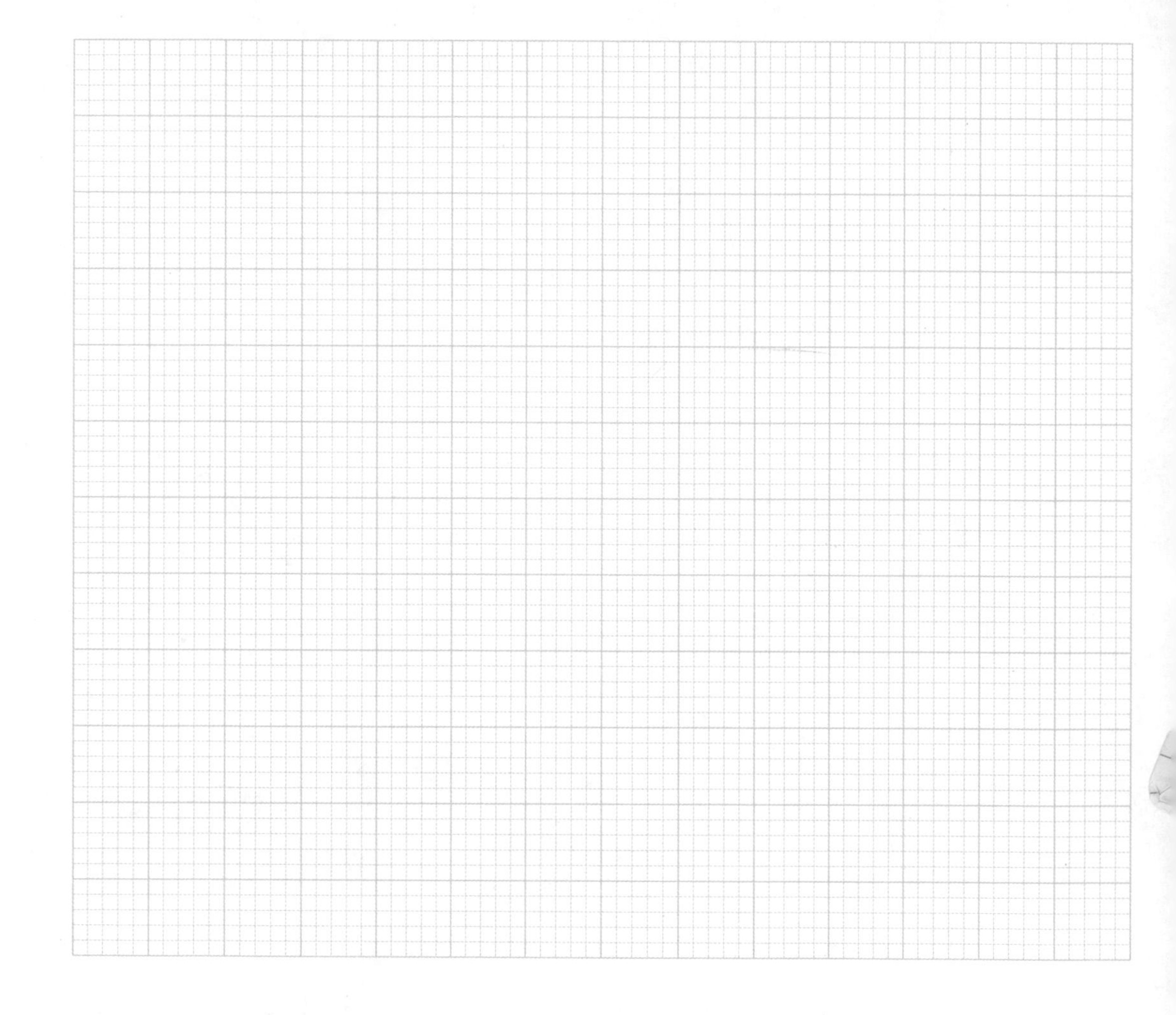

ANSWERS ON PAGE 94 ANSWERS ON PAGE 94 ANSWERS ON PAGE 94 ANSWERS ON PAGE 94

b) Complete the table below showing the 4-point moving average.

	Season	Total sold	Moving average
1998	Spring	43	
	Summer	17	
	Autumn	15	22.5
	Winter	15	23.5
1999	Spring	47	24.75
	Summer	19	25.5
	Autumn	18	28.0
	Winter	18	____
2000	Spring	57	____
	Summer	26	____
	Autumn	22	
	Winter	13	

③

c) Plot the 4-point moving averages on the graph. ③

d) Comment on the general trend and the seasonal variation.

....

....

.... ③

TOTAL 13

ANSWERS ON PAGE 94 ANSWERS ON PAGE 94 ANSWERS ON PAGE 94 ANSWERS ON PAGE 94

10 The table below shows the daily audiences for a production of 'An Inspector Calls' in a London theatre.

	Mon	Tues	Wed	Thurs	Fri	Sat
Week 1	250	318	330	446	570	610
Week 2	296	368	385	533	600	600
Week 3	240	335	325	500	565	583

a) Plot these figures on a graph. Use a scale of 1 cm for each day on the horizontal axis and 2 cm for 100 people on the vertical axis. (4)

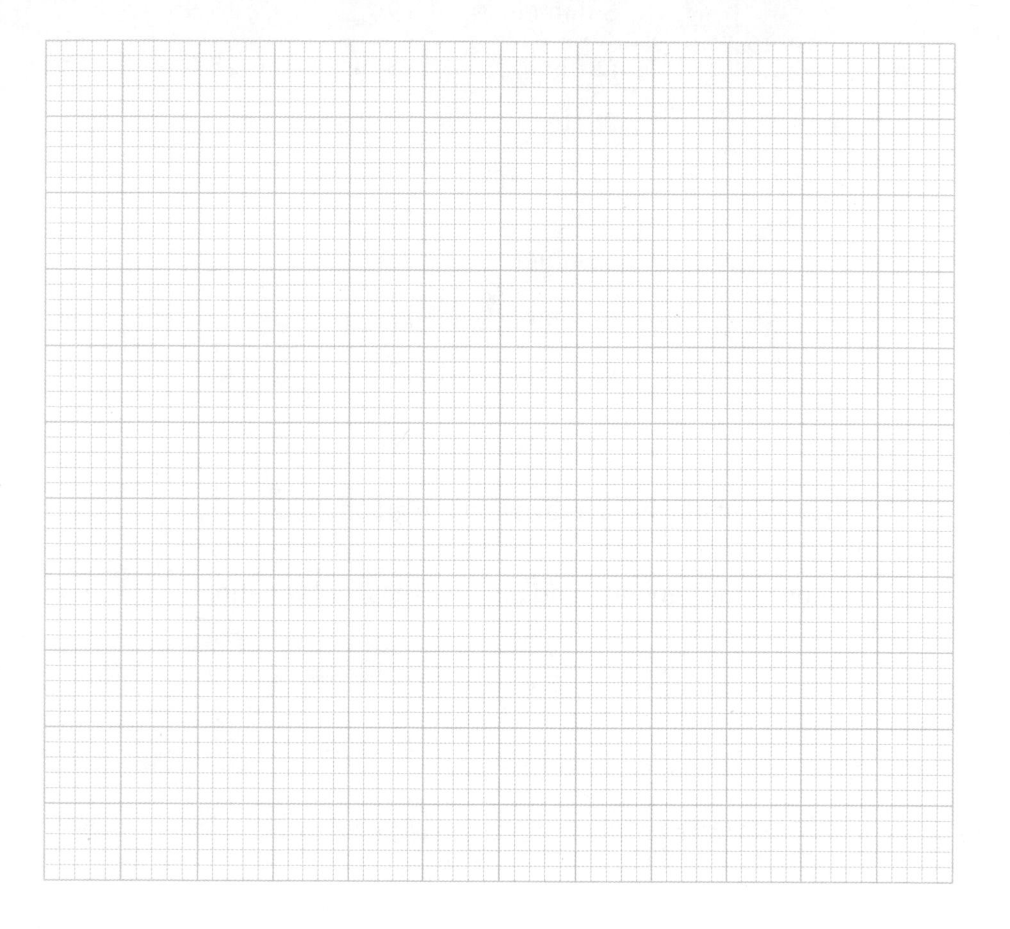

b) The table below has been used to calculate the 6-day moving averages, to the nearest whole number.

Calculate the last three entries for the table.

	Day	Audience	Moving average
Week 1	Monday	250	
	Tuesday	318	
	Wednesday	330	
	Thursday	446	421
	Friday	570	428
	Saturday	610	437
Week 2	Monday	296	446
	Tuesday	368	460
	Wednesday	385	465
	Thursday	533	464
	Friday	600	454
	Saturday	600	449
Week 3	Monday	240	439
	Tuesday	335	____
	Wednesday	325	____
	Thursday	500	____
	Friday	565	
	Saturday	583	

③

c) Plot the moving averages on the graph. ③

d) Comment on the general trend and the daily variation.

....

....

.... ③

TOTAL 13

ANSWERS ON PAGE 94 ANSWERS ON PAGE 94 ANSWERS ON PAGE 94 ANSWERS ON PAGE 94

11 As part of a survey into shopping in a rural area, 200 people were asked to record the distances, in kilometres, from their homes to the Post Office.

The results are shown in the table.

Distance (x km)	Frequency
$0 < x \leqslant 5$	24
$5 < x \leqslant 10$	46
$10 < x \leqslant 15$	68
$15 < x \leqslant 20$	38
$20 < x \leqslant 25$	16
$25 < x \leqslant 30$	8

a) Write down the modal class.

.... (1)

b) Calculate an estimate for the mean distance.

....

....

.... km (5)

c) Complete the cumulative frequency table.

Distance (in km), less than or equal to	Cumulative frequency
5	24
10	70

(3)

ANSWERS ON PAGE 94 ANSWERS ON PAGE 94 ANSWERS ON PAGE 94 ANSWERS ON PAGE 94

d) Using a scale of 2 cm for 5 km on the horizontal axis and 1 cm for 10 people on the vertical axis draw a cumulative frequency curve to show this data.

④

e) Use your graph to estimate:

i) the median distance .... km ②

ii) the interquartile range .... km ②

TOTAL 17

ANSWERS ON PAGE 94 ANSWERS ON PAGE 94 ANSWERS ON PAGE 94 ANSWERS ON PAGE 94

12 a) On his way to work, George has to pass two sets of traffic lights. They operate independently.

The probability that he has to stop at the first set is 0.3.
The probability that he has to stop at the second set is 0.6.

(i) Complete the tree diagram to show all the outcomes.

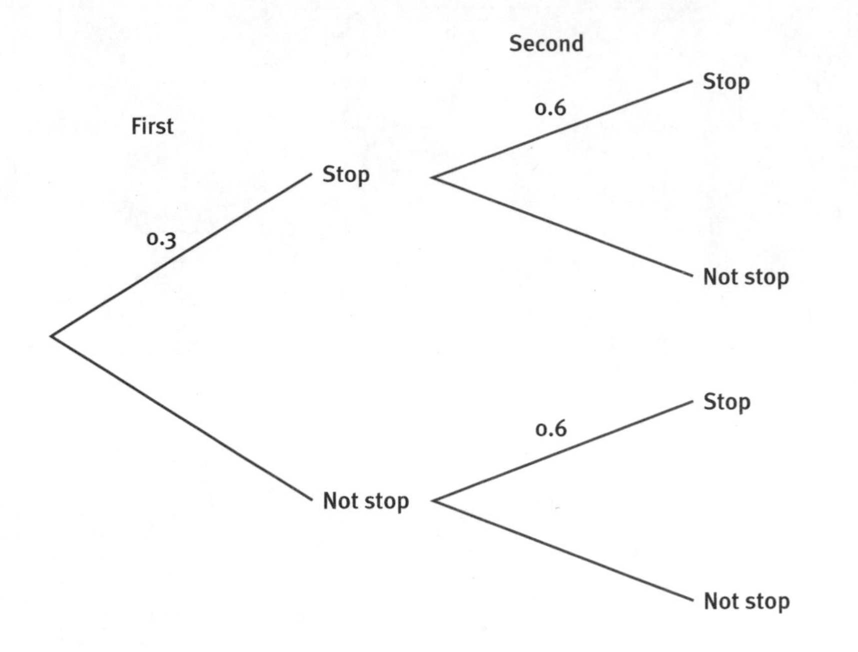

①

What is the probability

(ii) he does not have to stop?

.... ②

(iii) he has to stop at exactly one set?

....

....

.... ③

b) Traffic engineers connect the two sets of lights. This changes the probabilities.

The probability of not stopping at the first is 0.8.
If he does not stop at the first, the probability of not stopping at the second is 0.95.
If he does stop at the first, the probability of not stopping at the second is 0.35.

What is the probability

(i) he does not have to stop?

.... ③

(ii) he has to stop at exactly one set?

....

.... ③

TOTAL 12

ANSWERS ON PAGE 94 ANSWERS ON PAGE 94 ANSWERS ON PAGE 94 ANSWERS ON PAGE 94

13 The data shows the birth weights, in kilograms, of 20 babies.

2.4	2.7	3.2	2.5	3.1	1.9	3.5	4.0	2.7	4.0
4.1	2.8	4.2	3.4	1.6	4.2	3.8	2.6	3.0	2.4

a) Make a stem and leaf table, where 2 | 4 represents 2.4.

③

b) Find the median weight.

.... ①

TOTAL 4

---

14 A gardener measured the heights, in cm, of some trees four years after they had been planted. The results are shown below.

182	145	154	170	161	149	158	182
185	150	180	175	146	156	160	

a) Find the median and the quartiles.

....

....

.... ③

b) Draw a box and whisker plot.

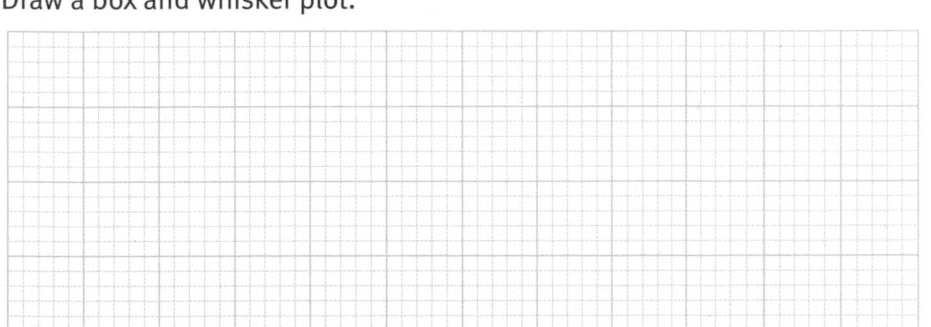

③

TOTAL 6

ANSWERS ON PAGE 94 ANSWERS ON PAGE 94 ANSWERS ON PAGE 94 ANSWERS ON PAGE 94

15 As part of his coursework Alan found the reaction times for some students in his year.

The stem and leaf table below shows the data he obtained:

	males		females
2		2	4 4
2	2 2 2 2 3 3	2	2 2 3 3 3 3
2	0 1	2	0 0 1 1
1	8 8 8 8 9	1	8
1	6 7 7	1	7 7 7 7
1	4 4 5	1	4 4
1		1	
1		1	
0	8	0	

0|8 means 8 hundreths of a second, i.e. 0.08 secs
so 1|4 means 14 hundreths of a second, 0.140 secs
and 2|4 means 24 hundreths of a second, i.e. 0.240 secs.

There were 20 males and 19 females in Alan's experiment.

a) Find the median time for

i) the males

.... ①

ii) the females.

.... ①

b) Find the lower and upper quartiles for

i) the males

....

.... ②

ii) the females.

....

.... ②

ANSWERS ON PAGE 94 ANSWERS ON PAGE 94 ANSWERS ON PAGE 94 ANSWERS ON PAGE 94

c) Make two comparisons about the data.

....

....

.... ②

TOTAL 8

---

16 Susan works for a train company. Her job is to analyse the numbers of passengers who travel on the company's trains.

Susan compares the number of passengers who travel on the 0800 train to London on Mondays and on Fridays over a period of a year.

The box plots show her data.

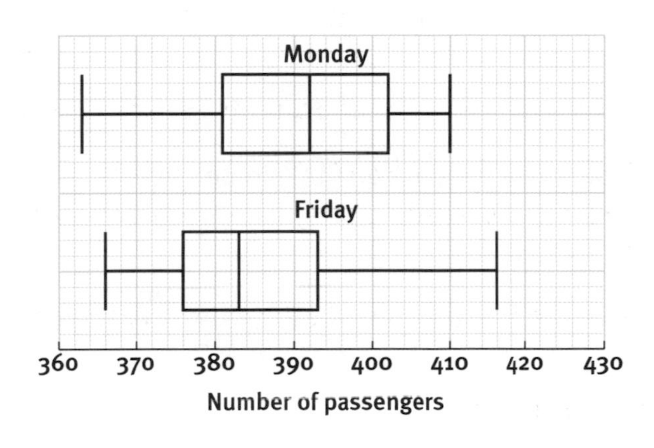

Make two comments about the data.

....

.... ②

TOTAL 2

ANSWERS ON PAGE 94 ANSWERS ON PAGE 94 ANSWERS ON PAGE 94 ANSWERS ON PAGE 94

As part of a primary school sports day 90 parents entered a three-legged egg and spoon race. The table summarises their times.

Time (t seconds)	Frequency
$10 \leq t < 20$	20
$20 \leq t < 30$	32
$30 \leq t < 40$	19
$40 \leq t < 60$	16
$60 \leq t < 90$	3
$t \geq 90$	0

a) Calculate an estimate of the mean time.

.... (4)

b) Which class contains the median?

.... (2)

c) Draw a histogram of the data on the grid below.

(4)

TOTAL 10

ANSWERS ON PAGE 94 ANSWERS ON PAGE 94 ANSWERS ON PAGE 94 ANSWERS ON PAGE 94

At Woodway Junior School 100 pupils took part in a challenge to build a large shape from different shaped and sized blocks.

Here are the times they took to complete the challenge.

Time (t seconds)	Number of pupils
$0 \leq t < 10$	0
$10 \leq t < 20$	9
$20 \leq t < 40$	20
$40 \leq t < 60$	30
$60 \leq t < 90$	15
$90 \leq t < 120$	6
$120 \leq t < 220$	20

a) Draw a histogram to show this information on the grid below. (4)

After further practice the same 100 pupils were given a similar challenge to complete. Their times are shown in the histogram below.

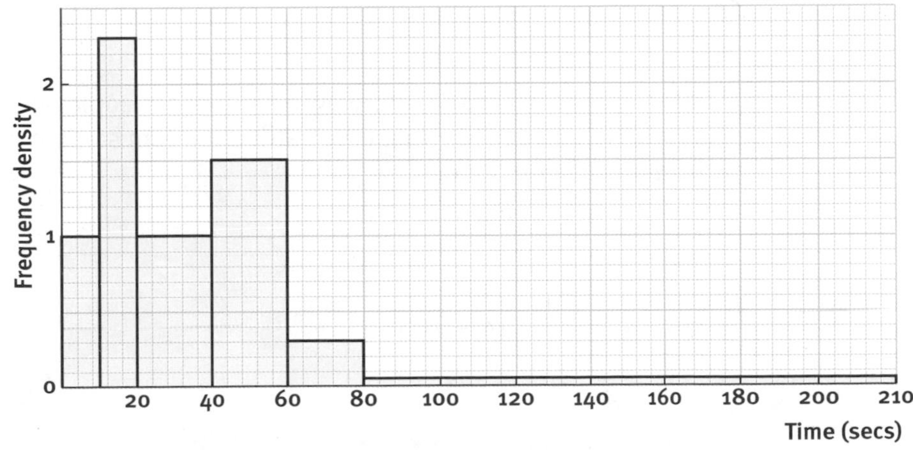

b) Make two comparisons about the distributions for the time taken to complete the challenge before and after practice.

.... (2)

TOTAL 6

19 A college runs evening classes in different crafts such as cookery, pottery, art and woodwork.

The people who attend these evening classes are grouped into ages as shown in the table.

Age (x years)	Number of people
$20 \leqslant x < 25$	6
$25 \leqslant x < 30$	7
$30 \leqslant x < 40$	59
$40 \leqslant x < 50$	62
$50 \leqslant x < 60$	16
$60 \leqslant x < 80$	10

a) Complete the histogram below to show this distribution.

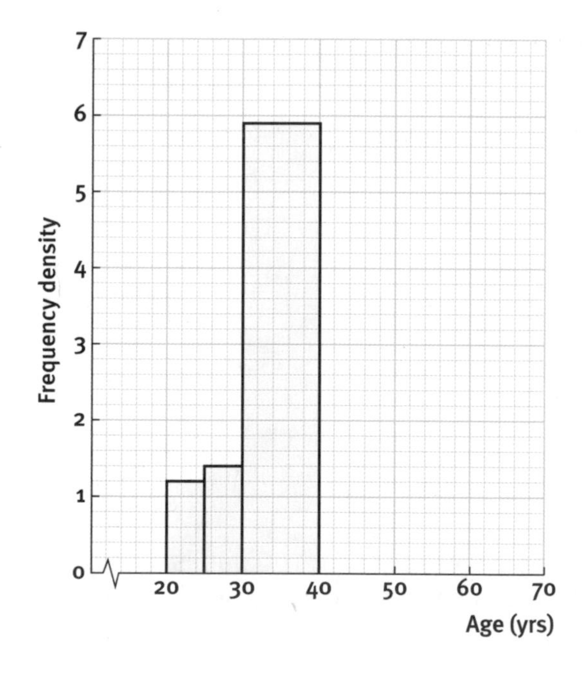

(5)

b) The principal of the college wishes to interview a sample of people who attend the evening classes. He decides to choose a stratifed random sample of 30 people. Show how this can be done.

....

....

....

....

.... (4)

TOTAL 9

ANSWERS ON PAGE 94 ANSWERS ON PAGE 94 ANSWERS ON PAGE 94 ANSWERS ON PAGE 94

20 A poultry farmer takes a sample of 100 eggs laid by his hens.

The masses of the eggs are given in the table.

Mass (m g)	Frequency	Frequency density
$40 \leq m < 42$	3	
$42 \leq m < 46$	7	
$46 \leq m < 54$	28	
$54 \leq m < 62$	36	
$62 \leq m < 75$	26	

a) Use the table to calculate an estimate of the mean weight of the eggs.

....

.... (4)

b) Calculate the frequency densities. (3)

c) On the grid below, draw a histogram to show this data.

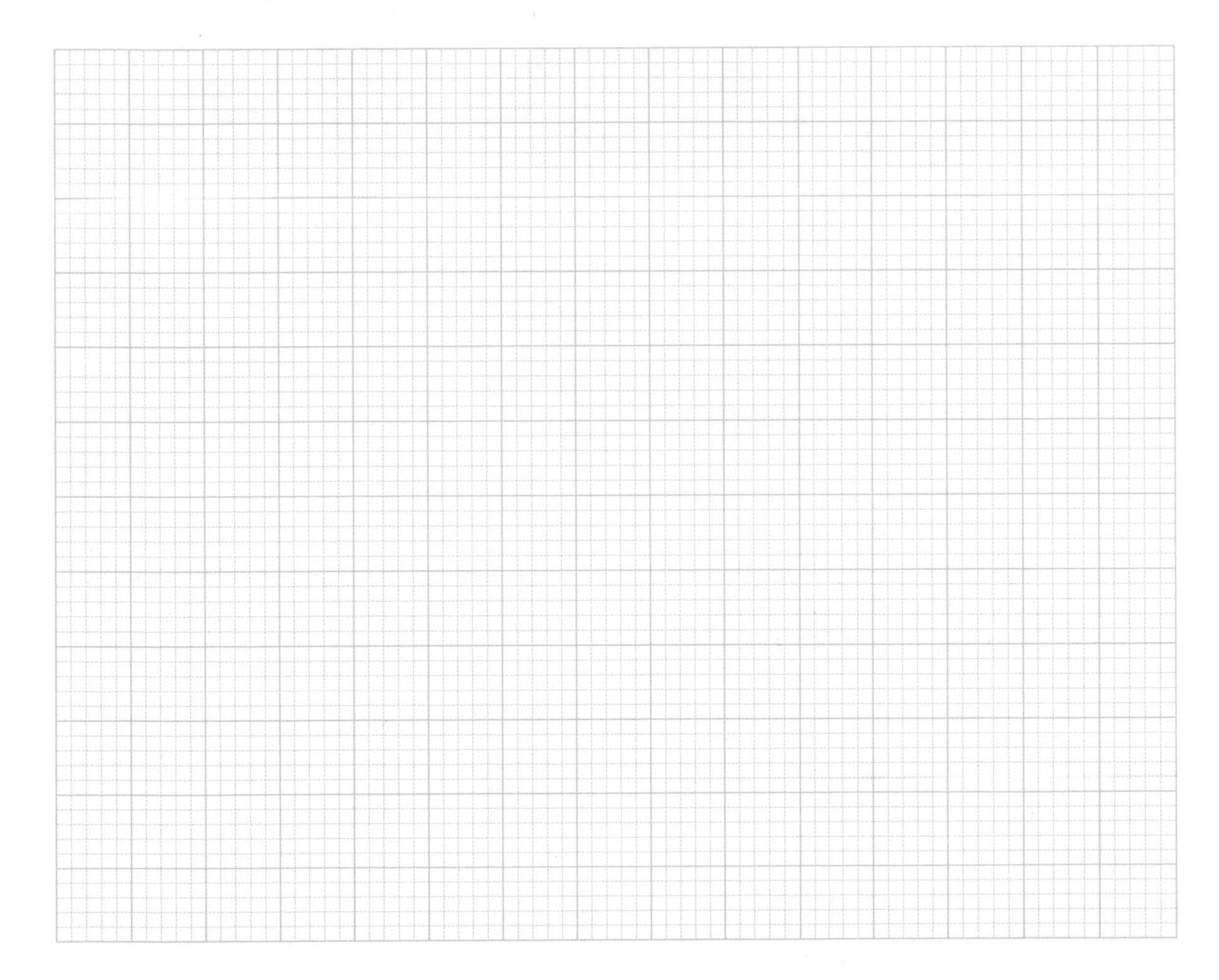

(3)

TOTAL 10

ANSWERS ON PAGE 94 ANSWERS ON PAGE 94 ANSWERS ON PAGE 94 ANSWERS ON PAGE 94

21 Three numbers are represented by a, b and c where $b = a + 3$ and $c = b + 6$.

a) Write c in terms of a.

.... ②

b) Find the median of the three numbers.

.... ②

c) Find the mean of the three numbers.

....

.... ②

d) What is the difference between the mean and the median?

.... ②

TOTAL 8

---

22 A supermarket manager compared the times customers had to wait at his supermarket, 'Johnson's Supermarket,' and at a rival supermarket, 'Foods@Us'.

The times, in seconds, are summarised in the table below.

	Median	Lower quartile	Upper quartile	Minimum	Maximum
Johnson's	41	35	55	20	70
Foods@Us	45	30	60	25	68

a) Draw two box plots on the grid below to show this data.

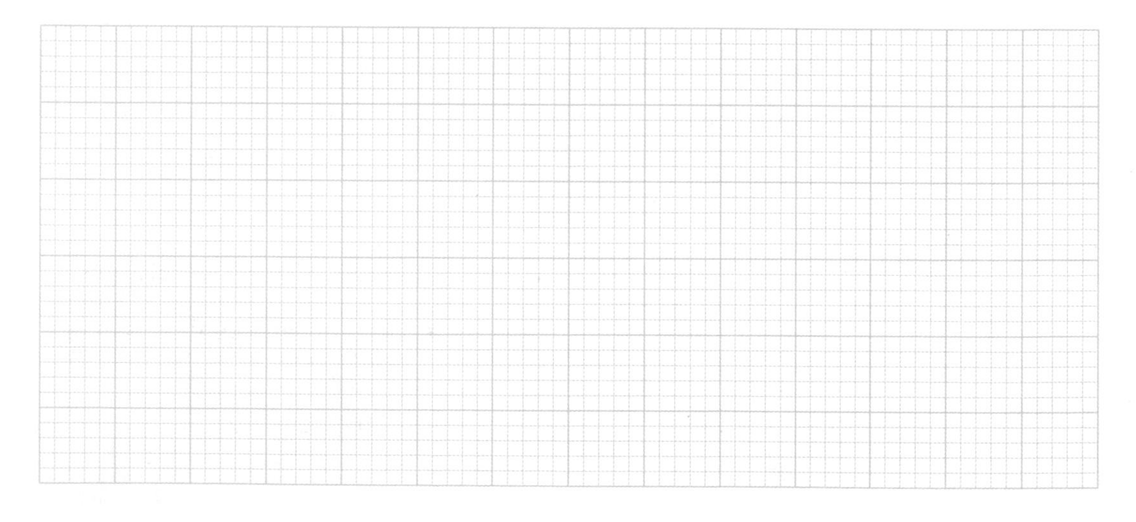

③

b) Use the information to make three comparisons between the two supermarkets.

....

.... ③

ANSWERS ON PAGE 94 ANSWERS ON PAGE 94 ANSWERS ON PAGE 94 ANSWERS ON PAGE 94

The mean waiting times for each supermarket are:

Johnson's 48 seconds *Foods@Us* 42 seconds

Here are two advertisements, one for each supermarket.

Johnson's	**Foods@Us**
The average waiting time at our supermarket tills is ____________	The average waiting time at our supermarket tills is ____________

c) Which average waiting time would you use in these advertisements. Give reasons for your answer.

....

.... (2)

TOTAL 8

23 A train operating company made a table showing how late 30 of its commuter trains were.

The data was available on this stem and leaf table. The times are in minutes.

2|4 means 24 minutes.

Stem	Leaf
1	1 3 5 6
2	4 6 8 8 9
3	5 5 6 7 8
4	0 1 1 2 2 2 3 4 5 6 7 8
5	0 0 1 4

The company claimed 'on average, our trains were less than 38 minutes late'.

A commuters' organisation disagreed. They claimed the average time was 42 minutes.

Comment on these claims explaining what these times represented.

....

.... (4)

TOTAL 4

24 a) This bar chart shows the results of a test taken by 30 students.

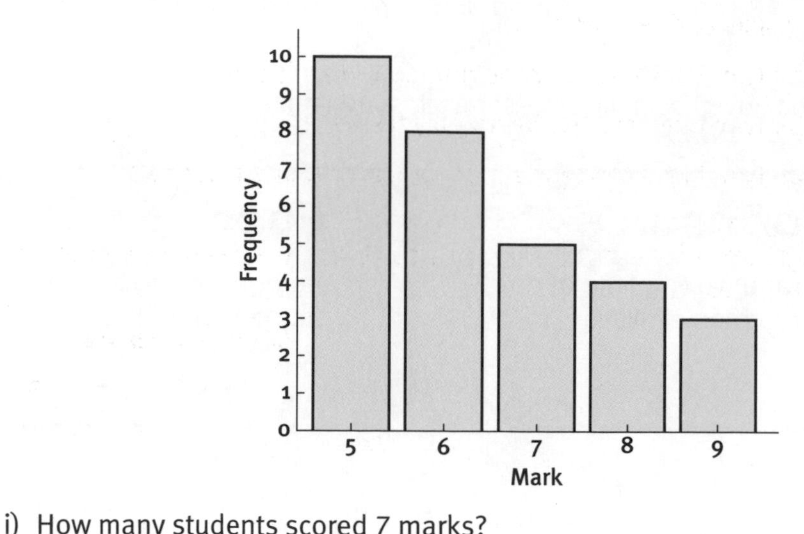

i) How many students scored 7 marks?

.... ①

ii) Use the bar chart to complete the frequency table. ②

Mark	5	6	7	8	9
Frequency					

iii) Write down the mode.

.... ①

iv) Find the median.

.... ①

v) Calculate the mean.

.... ③

vi) Sally draws a pie chart to show the same information.
Calculate the angle she should use to show the number of students who scored 6 marks.

....

.... ②

12 boys and 8 girls take another test.
The mean mark of these 20 students is 7.

b) i) Calculate the total marks of these 20 students.

.... ①

ii) The mean mark of the boys is 6.5.
Calculate the mean mark of the girls.

....

....

.... ③

TOTAL 14

Handling data

ANSWERS ON PAGE 94 ANSWERS ON PAGE 94 ANSWERS ON PAGE 94 ANSWERS ON PAGE 94

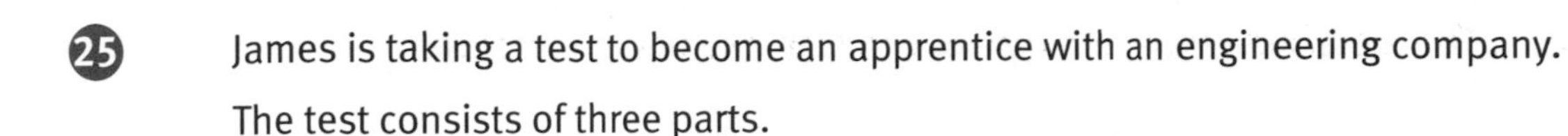

25 James is taking a test to become an apprentice with an engineering company.

The test consists of three parts.

The probability of James passing the numeracy test is 0.6.

The probability of him passing the literacy test is 0.7.

The probability of him passing the mechanical aptitude test is 0.4

a) Complete the tree diagram showing the possible outcomes of the three tests taken in the order given above.

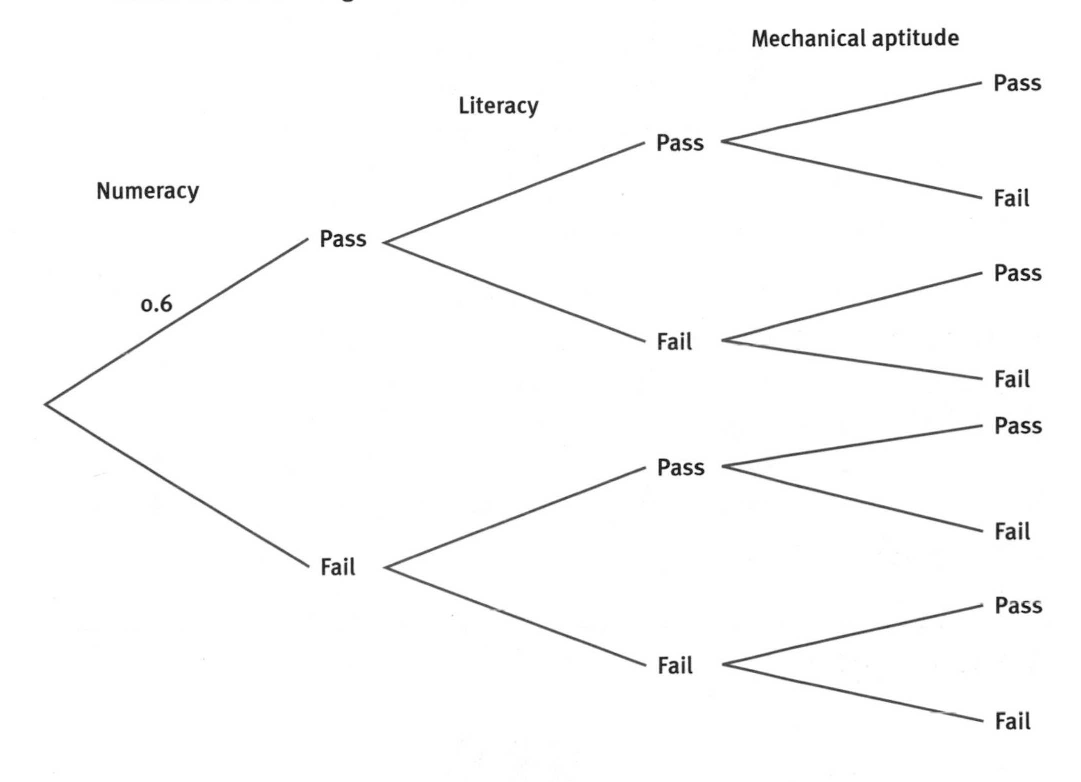

(5)

b) What is the probability that he will pass all three tests?

.... (3)

c) If James passes two of the tests, he is allowed to take the test he failed one more time.

What is the probability that he passes at least two of the tests?

.... (4)

d) If James passes fewer than two tests, he must repeat all three tests if he wishes to apply again. James is determined to become an apprentice.

What is the probability that he will have to repeat all three tests?

.... (2)

TOTAL 14

ANSWERS ON PAGE 94 ANSWERS ON PAGE 94 ANSWERS ON PAGE 94 ANSWERS ON PAGE 94

26 The table shows the numbers of boys and the numbers of girls in Year 7 and Year 8 of a school.

	Year 7	Year 8
Boys	100	50
Girls	90	60

The headteacher wants to find out what pupils think about the wearing of school uniform.

He selects two pupils at random to interview, one pupil to be chosen from Year 7 and one pupil from Year 8.

a) Calculate the probability that both pupils will be boys.

.... (2)

The headteacher decides to choose the 2 pupils at random from Year 7 and Year 8 together.

b) i) Calculate the probability that both pupils will be boys.

.... (2)

ii) Calculate the probability that both pupils will be from different Year groups.

.... (3)

TOTAL 7

27 Brian makes a 7-sided spinner as shown.

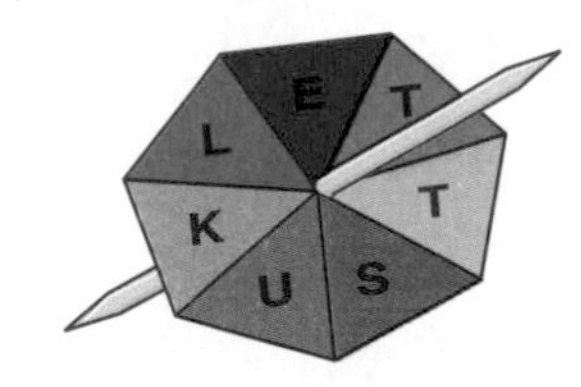

a) If Brian spins it once the probability that he

i) gets the letter E is .... (1)

ii) does not get the letter U is .... (1)

iii) gets a letter from the word SUCCESS is .... (2)

b) If he spins it 100 times about how many times should Brian expect to get the letter U or K?

....

.... (4)

Brian spins the spinner 700 times and his results are shown in the frequency table.

Result	L	E	T	S	U	K
Frequency	70	91	287	98	78	76

c) i) Do you think Brian's spinner is biased? Give a reason for your answer.

.... ②

ii) Find the relative frequency of the outcome being U or K.

.... ②

iii) The spinner is spun another 140 times. Predict the number of times the letter L is likely to occur.

.... ②

TOTAL 14

---

28 In a game, discs numbered from 1 to 100 are picked at random from a bag and not replaced. The numbers picked are crossed off on cards.

Bob and Sheila are playing with one card each.

No number appears on both cards.

2		33	50	71	
5	19	37		73	95
	26		65	79	
8	29	46		85	99

Bob's Card

11	30		74	91	
4		34	63	77	94
7	25	47		80	
9		49	69		98

Sheila's Card

a) What is the probability that the first number picked is:

i) one of the numbers on Bob's card?

.... ②

ii) one of the even numbers on Sheila's card?

.... ②

iii) one of the numbers on Bob's card or on Sheila's card?

.... ②

b) What is the probability that the first number picked is on Bob's card and the second number picked is on Sheila's card?

.... ②

c) What is the probability that the first two numbers picked are even numbers both on Bob's card **or** both on Sheila's card **or** one on each card?

.... ⑤

TOTAL 13

ANSWERS ON PAGE 94 ANSWERS ON PAGE 94 ANSWERS ON PAGE 94 ANSWERS ON PAGE 94

QUESTION BANK ANSWERS

1 a) systematic (1)
b) random (1)
c) quota (1)

2 a) 106 875 (2)
b) assume that the sample is representative (1)

EXAMINER'S TIP

Note that the question does not want to know the sampling method.

3 a) E.g. students might all come from the same area by bus, or they might belong to an athletics club and come for training (2)
b) i) not all students equally likely to be selected (2)
ii) not biased as position in alphabet is not important (1)
c) number students from 1 to 1350 and multiply each random number by 1000 or by 1350 to produce a 1-, 2- or 3-digit number (2)

4 All the bars in the sample were likely to have been made at the same time – therefore may not be representative. (2)

5 E.g.
- only those with phones contacted (1)
- excludes people from outside the town (1)
- excludes those who perhaps only shop there once every 2 weeks (1)

EXAMINER'S TIP

Note that there could be other valid reasons.

6 $\frac{100}{300} \times 50 = 17$ boys, $\frac{90}{300} \times 50 = 15$ girls (2)

EXAMINER'S TIP

The question refers to Year 7 pupils but the sample is to be taken from Years 7 and 8.

7 For Year 7 and Year 8 $\frac{140}{600} \times 30 = 7$ (2)

For Year 9 and Year 10 $\frac{100}{600} \times 30 = 5$ (1)

For Year 11 $\frac{120}{600} \times 30 = 6$ (1)

8 a) E.g. It would show the daily variation or trend over time (1)
b) moving averages are: £1167; £1184; £1207 (3)
c)

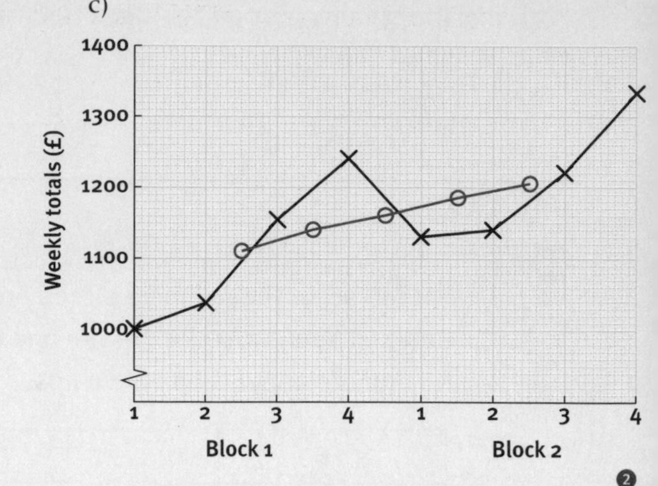

(2)

d) E.g.
- takings are lower in the first week of each block
- takings rise steadily in weeks 2–4
- general trend is upwards (3)

EXAMINER'S TIP

Remember to plot the moving averages at the midpoint of each time period.

9 a) see black graph line

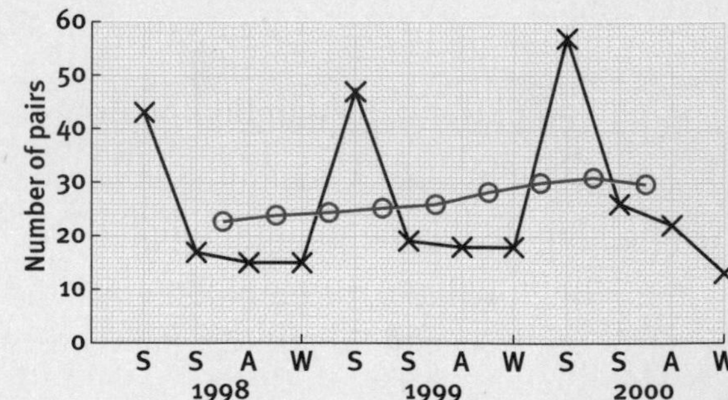

(4)

b) moving averages are: 29.75; 30.75; 29.5 (3)
c) see red graph line (3)
d) E.g.
- general trend is upwards – but a slight decrease at the end
- always much higher in the first quarter i.e. in spring (3)

EXAMINER'S TIP

Remember to plot the moving averages at the midpoint of each time period.

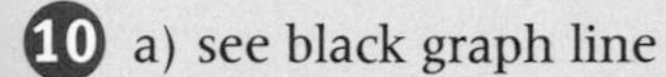

10 a) see black graph line

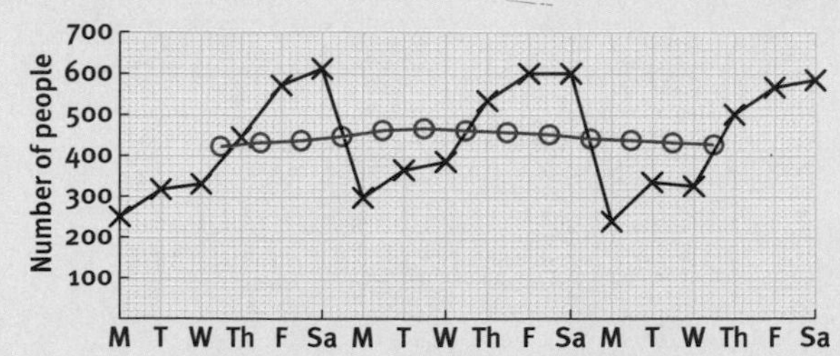

(4)

b) moving averages are 433, 428, 425 (3)
c) see red graph line (3)
d) E.g.
- general trend upwards initially then downwards
- Monday audience is always less than Friday or Saturday
- attendance rises through the week
- Saturday numbers always high (3)

EXAMINER'S TIP

Remember to plot the moving averages at the mid-point of each time period.

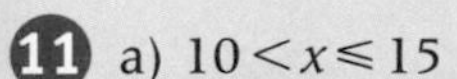

11 a) $10 < x \leq 15$ (1)
b) 12.5 km (5)
c)

Distance (in km), less than or equal to	Cumulative frequency
5	24
10	70
15	138
20	176
25	192
30	200

(3)

d)

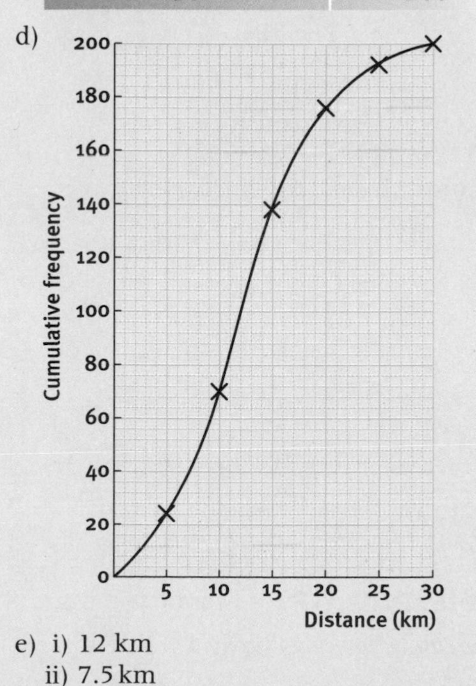

(4)

e) i) 12 km (2)
ii) 7.5 km (2)

EXAMINER'S TIP

Remember to use the mid-points of each group when calculating the mean distance.

12 a) i) Lower branches: 0.7, 0.4, 0.4 (1)
ii) 0.28 (2)
iii) $0.12 + 0.42 = 0.54$ (3)
b) i) 0.76 (3)
ii) $0.13 + 0.04 = 0.17$ (3)

EXAMINER'S TIP

You may find a tree diagram useful in part (b) as well. Take care to get the Stop/Not stop the right way round.

13 a)

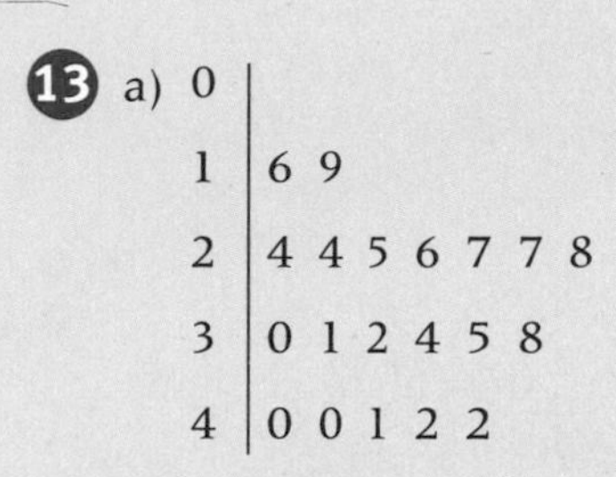

Stem	Leaf
0	
1	6 9
2	4 4 5 6 7 7 8
3	0 1 2 4 5 8
4	0 0 1 2 2

(3)

b) 3.05 kg (1)

EXAMINER'S TIP

Remember to order the values along each row.

14 a) Median = 160 (1)
lower quartile = 150 (1)
upper quartile = 180 (1)
b)

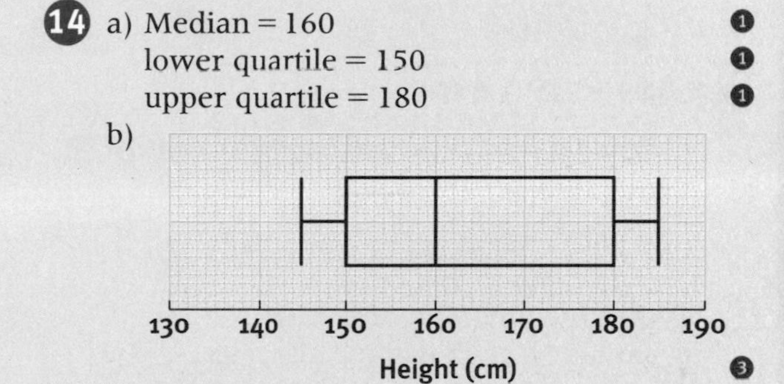

(3)

EXAMINER'S TIP

Remember to order the values along each row.

15 a) Median times
i) male = 18 hundreths i.e. 0.18 secs (1)
ii) female = 21 hundreths, i.e. 0.21 secs (1)
b) i) Males – upper quartile = 0.23 secs, lower quartile = 0.165 secs (2)
ii) Females – upper quartile = 0.23 secs, lower quartile = 0.17 secs (2)
c) E.g.
- males times are more spread out – the range is greater
- males generally have quicker reaction times (2)

EXAMINER'S TIP

Remember to order the values along each row.

FOR MORE INFORMATION ON THIS TOPIC ... SEE REVISE GCSE MATHS ... CHAPTER 4

16 E.g.

- The range for Monday (410 − 363) is less than for Friday (416 − 366).
- The interquartile range for Monday is greater than for Friday.
- The median number is greater on Monday.
- 75% of the trains on Monday carry over 381 passengers while on Friday 75% of the trains carry over 376 passengers. ❷

EXAMINER'S TIP

Remember that each section of the box plot represents 25% of the data.

17 a) 31 s ❹

b) $20 \leqslant t < 30$ ❷

c) Check your graph – the column heights should be as shown in this table:

Time (t seconds)	Height
$10 \leqslant t < 20$	2
$20 \leqslant t < 30$	3.2
$30 \leqslant t < 40$	1.9
$40 \leqslant t < 60$	0.8
$60 \leqslant t < 90$	0.1
$t \geqslant 90$	0

❹

EXAMINER'S TIP

Remember to calculate the frequency density.

18 a) Check your graph, the column heights should be as shown in this table:

Time (t seconds)	Height
$0 \leqslant t < 10$	0
$10 \leqslant t < 20$	0.9
$20 \leqslant t < 40$	1
$40 \leqslant t < 60$	1.5
$60 \leqslant t < 90$	0.5
$90 \leqslant t < 120$	0.2
$120 \leqslant t < 220$	0.2

❹

b) E.g.

- wider range of times after practice
- times decreased after practice
- the distribution is skewed to the left after practice
- the modal time is decreased ❷

EXAMINER'S TIP

Remember to calculate the frequency density.

19 a) The frequency densities are:

Age (x years)	Frequency density
$20 \leqslant x < 25$	1.2
$25 \leqslant x < 30$	1.4
$30 \leqslant x < 40$	5.9
$40 \leqslant x < 50$	6.2
$50 \leqslant x < 60$	1.6
$60 \leqslant x < 80$	0.5

❸

check your graph; if correct then ❷

b)

Age (x years)	Sample
$20 \leqslant x < 25$	1
$25 \leqslant x < 30$	1
$30 \leqslant x < 40$	11
$40 \leqslant x < 50$	12
$50 \leqslant x < 60$	3
$60 \leqslant x < 80$	2

❹

EXAMINER'S TIP

Part b) doesn't depend on a correct answer to part a).

20 a) 57 g ❹

b)

Mass (m g)	Frequency density
$40 \leqslant m < 42$	1.5
$42 \leqslant m < 46$	1.75
$46 \leqslant m < 54$	3.5
$54 \leqslant m < 62$	4.5
$62 \leqslant m < 75$	2

❸

c) Correct histogram: check heights = frequency densities above ❸

21 a) $c = (a + 3) + 6 = a + 9$ ❷

b) $a + 3$ ❷

c) Total $= a + (a + 3) + (a + 9) = 3a + 12$.

Mean $= \dfrac{3a + 12}{3} = a + 4$ ❷

d) 1 ❷

EXAMINER'S TIP

Take care when dividing by 3 in part c).

FOR MORE INFORMATION ON THIS TOPIC ... SEE REVISE GCSE MATHS ... CHAPTER 4

22 a)

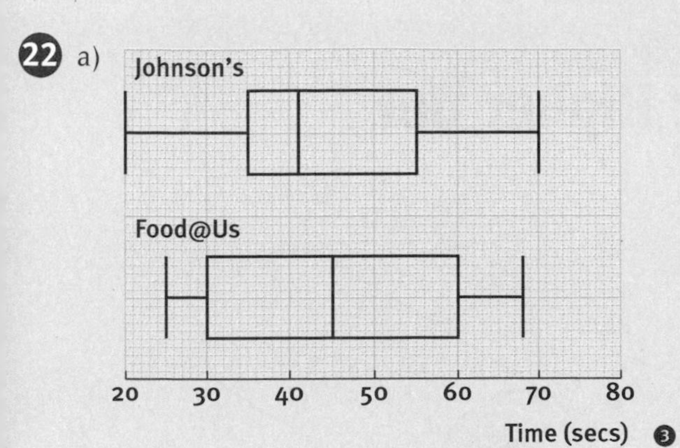

❸

b) E.g.

- The interquartile range for Johnson's is less.
- 50% of customers wait 41 seconds or less at Johnson's compared with 45 seconds or less at Foods@Us.
- The median for Johnson's is less than for Foods@Us.
- The range for Johnson's is greater than the range for Foods@Us.

[❸ for any three from the list]

c) Johnson's should use the median
Foods@Us should use the mean ❷

EXAMINER'S TIP

Remember that each section of the box plot represents 25% of the data.

23 Median = 40.5 minutes; Mean = 36.6 minutes; Mode = 42 minutes. E.g. the company are correct if they use and state that they are using the mean; the commuters are correct if they quote the mode – the most common. [❷ for calculating the mean, ❶ for finding the mode, ❶ for the comment]

24

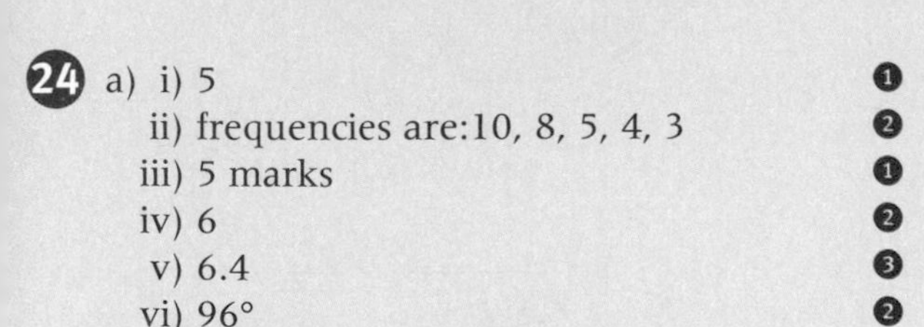

a) i) 5 ❶
ii) frequencies are:10, 8, 5, 4, 3 ❷
iii) 5 marks ❶
iv) 6 ❷
v) 6.4 ❸
vi) 96° ❷

b) i) 140 ❶ ii) 7.75 ❸

EXAMINER'S TIP

Use the answer for b) i) and the total mark for the boys to calculate b) ii)

25 a)

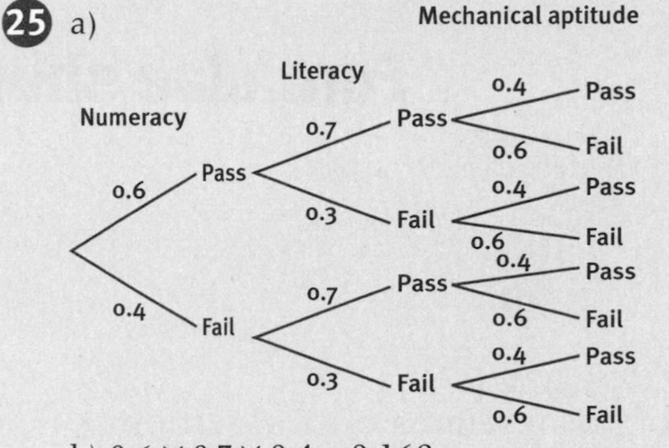

❺

b) $0.6 \times 0.7 \times 0.4 = 0.168$ ❸

c) PPP + PPF + PFP + FPP $= 0.168 + 0.252 + 0.072 + 0.112 = 0.604$ ❹

d) $1 - 0.604 = 0.396$ ❷

EXAMINER'S TIP

In part c) be systematic so you don't miss a case.

26

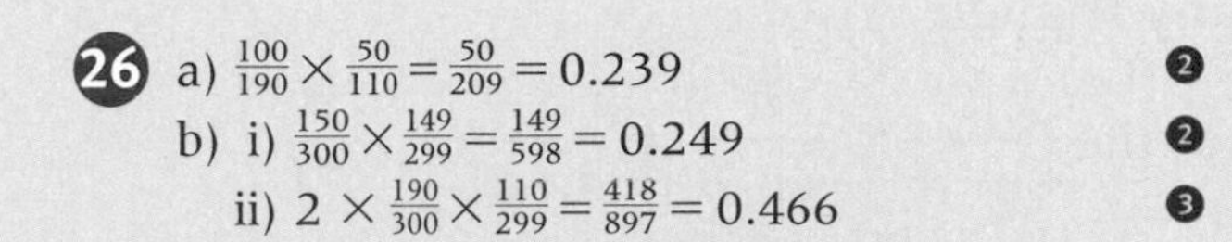

a) $\frac{100}{190} \times \frac{50}{110} = \frac{50}{209} = 0.239$ ❷

b) i) $\frac{150}{300} \times \frac{149}{299} = \frac{149}{598} = 0.249$ ❷

ii) $2 \times \frac{190}{300} \times \frac{110}{299} = \frac{418}{897} = 0.466$ ❸

EXAMINER'S TIP

In part b) i) and ii) remember that the second probability involves one less in the total.

27

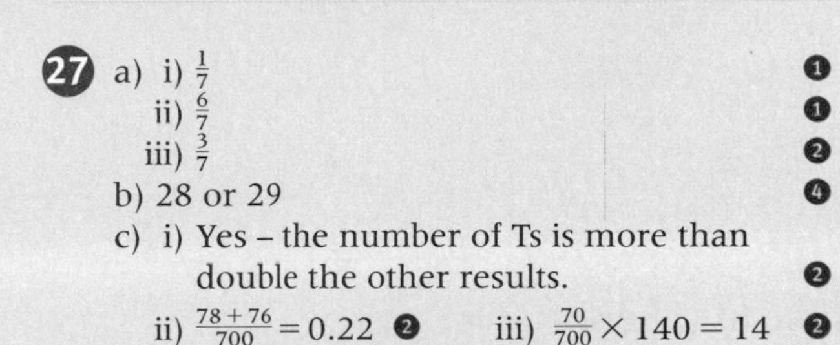

a) i) $\frac{1}{7}$ ❶
ii) $\frac{6}{7}$ ❶
iii) $\frac{3}{7}$ ❷

b) 28 or 29 ❹

c) i) Yes – the number of Ts is more than double the other results. ❷

ii) $\frac{78+76}{700} = 0.22$ ❷ iii) $\frac{70}{700} \times 140 = 14$ ❷

EXAMINER'S TIP

In part c) i) remember to give reasons for your answer.

28

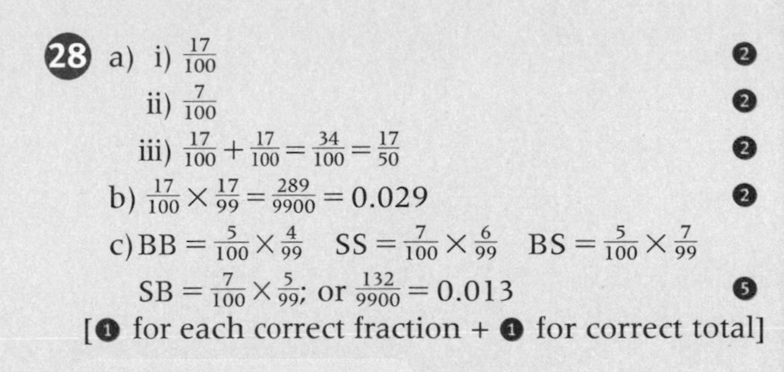

a) i) $\frac{17}{100}$ ❷
ii) $\frac{7}{100}$ ❷
iii) $\frac{17}{100} + \frac{17}{100} = \frac{34}{100} = \frac{17}{50}$ ❷

b) $\frac{17}{100} \times \frac{17}{99} = \frac{289}{9900} = 0.029$ ❷

c) $BB = \frac{5}{100} \times \frac{4}{99}$ $SS = \frac{7}{100} \times \frac{6}{99}$ $BS = \frac{5}{100} \times \frac{7}{99}$
$SB = \frac{7}{100} \times \frac{5}{99}$; or $\frac{132}{9900} = 0.013$ ❺

[❶ for each correct fraction + ❶ for correct total]

EXAMINER'S TIP

In part c) be systematic – as the answer indicates.

FOR MORE INFORMATION ON THIS TOPIC ... SEE REVISE GCSE MATHS ... CHAPTER 4

Formulae sheet: Higher tier

Volume of prism = (area of cross section) × length

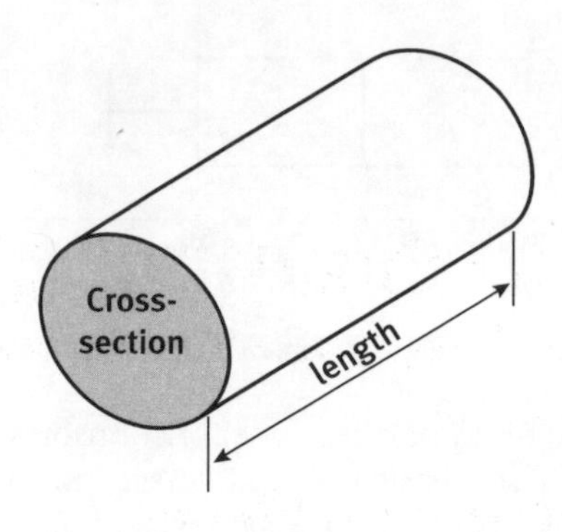

In any triangle ABC

Sine rule $\dfrac{a}{\sin A} = \dfrac{b}{\sin B} = \dfrac{c}{\sin C}$

Cosine rule $a^2 = b^2 + c^2 - 2bc \cos A$

Area of triangle $= \dfrac{1}{2}ab \sin C$

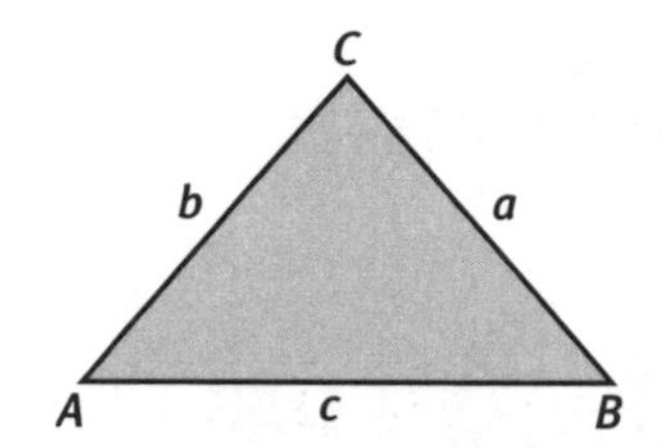

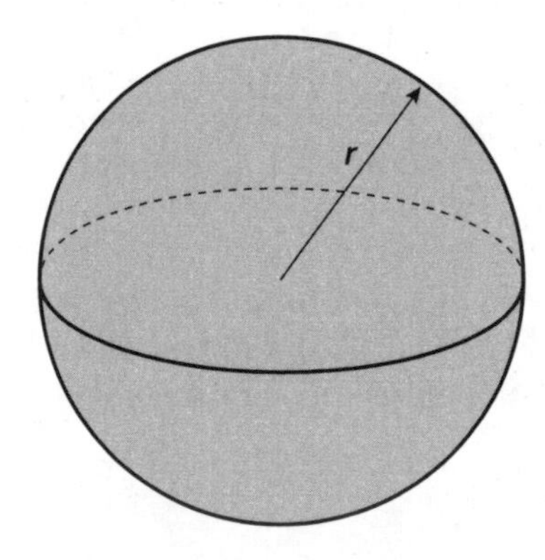

Volume of sphere $= \dfrac{4}{3}\pi r^3$

Surface area of sphere $= 4\pi r^2$

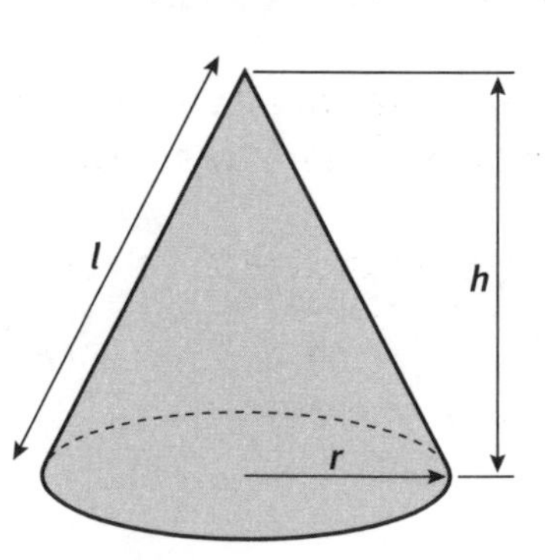

Volume of cone $= \dfrac{1}{3}\pi r^2 h$

Curved surface area of cone $= \pi r l$

The Quadratic Equation

The solution of $ax^2 = bx + c = 0$ where $a \neq 0$, area given by $x = \dfrac{-b \pm \sqrt{(b^2 - 4ac)}}{2a}$

Centre number
Candidate number
Surname and initials

Mathematics
Paper 1

Higher tier

Time: two hours

Instructions to candidates

Write your name, centre number and candidate number in the boxes at the top of this page.

Answer ALL questions in the spaces provided on the question paper.

Show all stages in any calculations and state the units. **You must not use a calculator in this paper**.

Include diagrams in your answers where this may be helpful.

Information for candidates

The number of marks available is given in brackets **[2]** at the end of each question or part question.

The marks allocated and the spaces provided for your answers are a good indication of the length of answer required.

For Examiner's use only	
1	
2	
3	
4	
5	
6	
7	
8	
9	
10	
11	
12	
13	
14	
15	
16	
Total	

Leave blank

1 **(a)** Write 30 as a product of prime factors.

....

.... **[2]**

(b) Find **(i)** the LCM **(ii)** the HCF of 30 and 54.

....

....

.... **[2]**

(Total 4 marks)

2 Mark and Brian drove to Cambridge.
Mark took $2\frac{3}{4}$ hours. Brian's journey was $3\frac{1}{6}$ hours.

Work out

(a) $2\frac{3}{4} + 3\frac{1}{6}$

....

....

.... **[3]**

(b) $3\frac{1}{6} - 2\frac{3}{4}$

....

.... **[3]**

(Total 6 marks)

3 In a sale, prices were reduced by 20%.
A suit in the sale cost £160.

Calculate the original price.

SALE
Everything reduced
20%

....

....

.... **[3]**

(Total 3 marks)

Leave blank

4 In each case, make c the subject of the formula.

(a) $y = mx + c$

....

.... **[1]**

(b) $e = mc^2$

....

....

.... **[2]**

(c) $a = \dfrac{bc - 4}{2c}$

....

....

.... **[2]**

(Total 5 marks)

5 In triangle ABC, DE is parallel to BC. AE = ED = DB = 3 cm, BC = 9 cm.

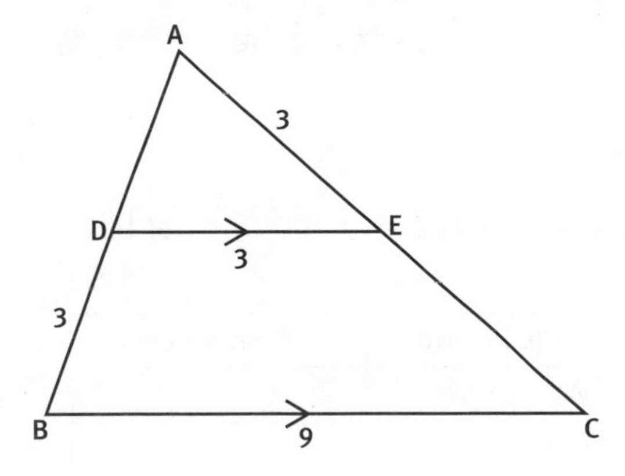

Find the lengths of:

(a) EC

....

....

.... **[2]**

(b) AD.

....

....

.... **[2]**

(Total 4 marks)

[turn over

6 A test is conducted to find out how long torch batteries last.
A sample of 100 *Britelite* batteries is tested. The results are shown on the cumulative frequency diagram.

Leave blar

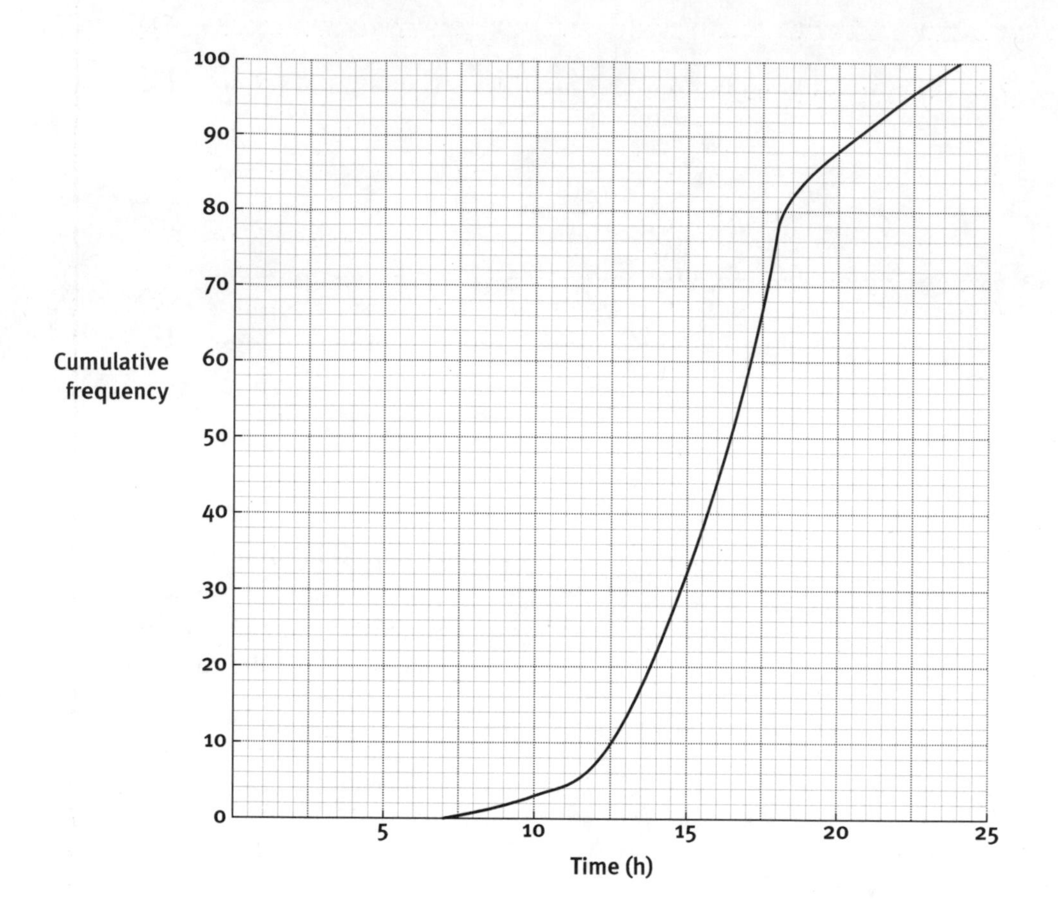

This table shows the results for a sample of 100 *Durabat* batteries.

Life (L hours)	Frequency
At least 10	1
$10 < L \leqslant 12$	5
$12 < L \leqslant 14$	27
$14 < L \leqslant 16$	42
$16 < L \leqslant 18$	20
$18 < L \leqslant 20$	5

Leave blank

(a) Draw a cumulative frequency diagram for *Durabat* on the same axes.

....

.... **[4]**

(b) Draw box plots for both samples.

....

....

.... **[5]**

(c) Compare the results for the two types of battery.

....

....

.... **[1]**

(Total 10 marks)

Leave blan

7 **(a)** Express $2x^2 + 5x - 3$ in factors.

....

....

.... **[2]**

(b) Hence solve

(i) $2x^2 + 5x - 3 = 0$

....

....

.... **[1]**

(ii) $2x^2 + 5x - 3 \leqslant 0$

....

....

.... **[2]**

(Total 5 marks)

8 **(a)** Solve this equation.

$$\frac{3x}{4} = 12$$

....

....

.... **[2]**

(b) Solve these simultaneous equations algebraically.

$$2x + 3y = -3$$
$$3x - 2y = 8\tfrac{1}{2}$$

....

....

.... **[4]**

(Total 6 marks)

Leave blank

9 ABCD is a parallelogram

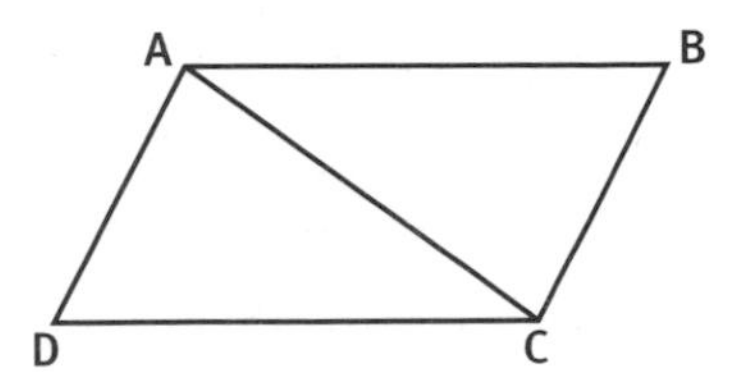

(a) Prove that triangle ACD is congruent to triangle CAB.

....

....

.... **[3]**

(b) Describe a transformation that could also be used to demonstrate that these two triangles are congruent.

....

.... **[3]**

(Total 6 marks)

10 Two sets of traffic lights are connected. The first set is green when you arrive with probability 0.6. If the first set is green the probability that the second set will also be green when you arrive is 0.8.

If the first set was not green and you had to stop, the probability that the second set will not be green when you arrive is 0.7.

Mr Parsons drives through the traffic lights.

What is the probability that he has to stop exactly once?

....

....

....

....

....

.... **[4]**

(Total 4 marks)

11 Simplify these expressions and write as surds in the form $a + b\sqrt{c}$ or $b\sqrt{c}$.

Leave blank

(a) $\sqrt{10} \times \sqrt{50}$

.... **[1]**

(b) $\sqrt{5}(1 + \sqrt{5})$

.... **[1]**

(c) $\dfrac{1 + \sqrt{5}}{1 - \sqrt{5}}$

....

....

.... **[3]**

(Total 5 marks)

12 This is the graph of

$$y = x^3 - 2x^2 + x - 1.$$

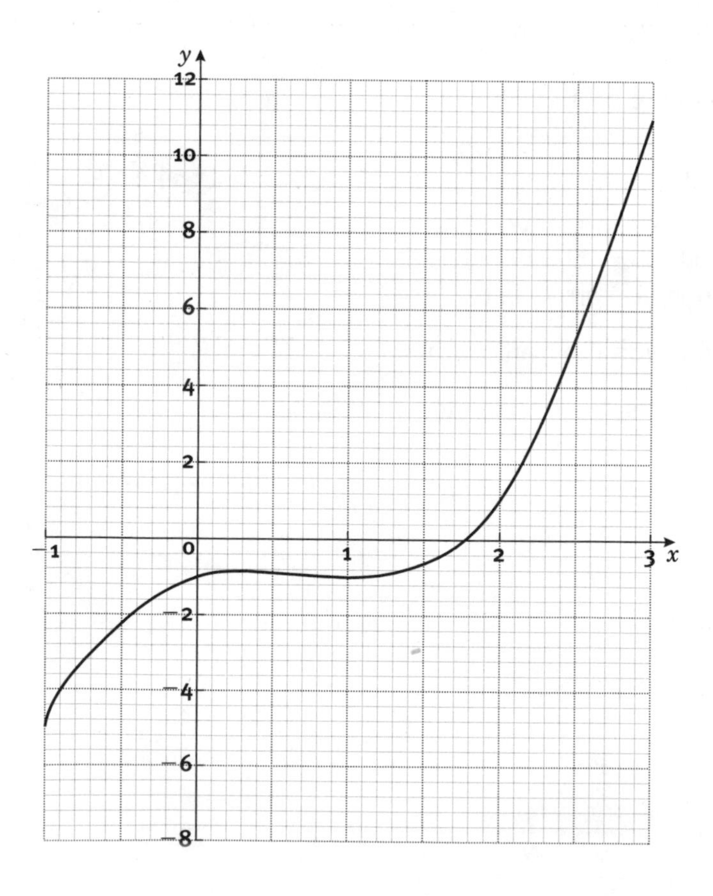

Use the graph to solve the following equations.

Leave blank

(a) $x^3 - 2x^2 + x - 1 = 0$

.... **[1]**

(b) $x^3 - 2x^2 + x = 0$

....

....

.... **[2]**

(c) $x^3 - 2x^2 + 2x - 2 = 0$

....

....

.... **[3]**

(d) $x^3 - 2x^2 + 1 = 0$

....

.... **[3]**

(e) Are there any more roots of the equation in part **(d)**? Explain.

....

.... **[2]**

(Total 11 marks)

[turn over

Leave blank

13 **(a)** Write as single fractions in their simplest forms.

(i) $\frac{x}{2} - \frac{2x + 1}{3}$

....

.... **[2]**

(ii) $\frac{2}{x} - \frac{3}{2x + 1}$

....

....

....

.... **[4]**

(b) Simplify these expressions.

(i) $\frac{x^2 - x}{x^2 - 1}$

....

....

.... **[3]**

(ii) $\frac{5a^2b \times b^3}{10a^3b^2}$

....

....

.... **[3]**

(Total 12 marks)

14 O is the centre of the circle through A, B and C.
TCS is the tangent to the circle at C.

Angle AOC = 124°.

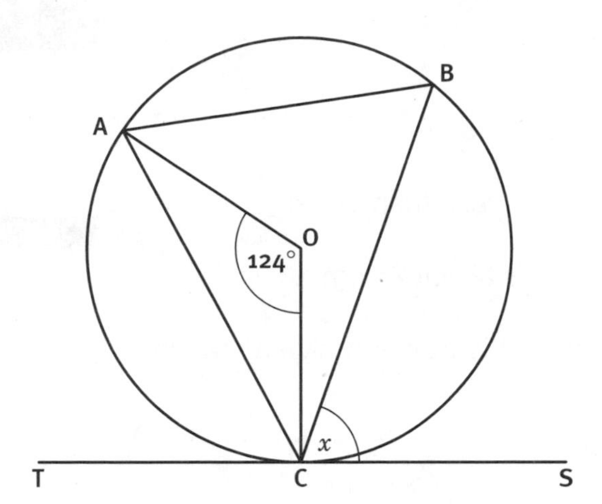

(a) Find the size of angle ACT.
Give a reason for each step.

....

....

....

.... **[3]**

(b) Express the angles of triangle ABC in terms of x.

....

....

....

.... **[3]**

(Total 6 marks)

Leave blank

[turn over

15 This wine bottle contains 750 ml.
A mathematically similar bottle (half-bottle) contains 375 ml.

The height of the 750 ml bottle is 32 cm.

Here are three statements.

(a) The height of the half-bottle is 16 cm.

(b) The height of the half-bottle is less than 16 cm.

(c) The height of the half-bottle is about 25 cm.

For each statement, say whether it is true and give a reason.

....

....

....

....

....

....

....

.... **[6]**

(Total 6 marks)

Leave blar

Leave blank

16 **(a)** On the diagram draw vectors **c** and **d**, where

$$\mathbf{c} = \mathbf{a} + 2\mathbf{b}, \quad \mathbf{d} = 2\mathbf{a} + \mathbf{b}.$$

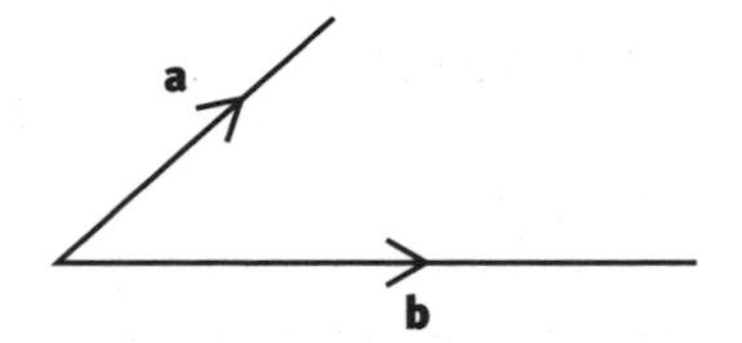

[2]

The points A, B, C and D have position vectors **a**, **b**, **c** and **d**.

(b) Find $\overrightarrow{AD}$ in terms of **a** and **b**.

....

.... **[2]**

(c) Hence prove that ABCD is a parallelogram.

....

....

....

....

....

.... **[3]**

(Total 7 marks)

Centre number
Candidate number
Surname and initials

Mathematics
Paper 2

Higher tier

Time: two hours

Instructions to candidates

Write your name, centre number and candidate number in the boxes at the top of this page.

Answer ALL questions in the spaces provided on the question paper.

Show all stages in any calculations and state the units. You are expected to use a calculator in this paper.

Include diagrams in your answers where this may be helpful.

Information for candidates

The number of marks available is given in brackets **[2]** at the end of each question or part question.

The marks allocated and the spaces provided for your answers are a good indication of the length of answer required.

For Examiner's use only	
1	
2	
3	
4	
5	
6	
7	
8	
9	
10	
11	
12	
13	
14	
15	
16	
Total	

Leave blank

1 Use a trial and improvement method to find the value of x, correct to two decimal places, when

$$x^3 + x - 1 = 0$$

....

....

....

....

....

.... **[4]**

(Total 4 marks)

2 **(a)** Use your calculator to find the values of the following expressions.
Give your answers correct to two decimal places.

(i) $\dfrac{429 \times 0.93}{0.33 + 0.54}$

....

.... **[2]**

(ii) $\sqrt{0.63^3 + 0.63^2}$

....

.... **[2]**

(b) The distances of three planets from the Sun are:

Earth	1.5×10^8 km
Pluto	6.02×10^9 km
Mercury	5.8×10^7 km

The ratio:

Pluto's distance from the Sun : Earth's distance from the Sun $= 4.01 \times 10^1 : 1$

What is the corresponding ratio for Mercury and Earth?
Give your answer in standard form correct to three significant figures.

....

.... **[2]**

(Total 6 marks)

Leave blank

3 **(a)** Simplify

$$3(5m + 4) - 4(3m + 5)$$

.... **[2]**

(b) **(i)** Factorise

$$x^2 - 3x - 4$$

....

.... **[2]**

(ii) Solve this equation.

$$x^2 - 3x - 4 = 0$$

.... **[1]**

(Total 5 marks)

4 In 2001 Bob was quoted a basic car insurance premium of £1100 for one year.

Bob agreed to pay the first £500 of any claim and this meant he gained a 15% discount.

Because Bob had a 'no claim' bonus he also received a further 20% discount on the reduced premium.

(a) How much did Bob have to pay?

....

.... **[2]**

Bob decided to pay by 12 equal monthly instalments, but this added 10% to the reduced premium.

(b) How much did Bob pay each month?

....

.... **[2]**

For a different make of car, in 2002, Bob is quoted a basic insurance premium of £1500.

This was 15% more than the 2001 basic premium on this make of car.

(c) Find the value of the 2001 basic premium for this car. Round your answer to the nearest pound.

....

.... **[3]**

(Total 7 marks)

5 **(a)** Bill and Tony work in the same office. They have a wager that they will come to work in the same colour of shirt one day.

Bill has 10 shirts: 4 white, 3 blue, 3 grey

Tony has 8 shirts: 2 white, 2 blue, 4 grey

What is the probability that they will wear the same colour shirt?

....

.... **[5]**

(b) Mike wears a clean shirt every day.

He also has 10 shirts: 6 white, 2 blue, 2 grey

Bill and Tony also wager that Mike will wear the same colour shirt for 3 consecutive days.

What is the probability that this will happen?

....

.... **[4]**

(Total 9 marks)

Leave blank

Leave blank

6 The diagram shows a solid bar of lead.
The bar is a prism. The cross-section of the prism is an isosceles trapezium.

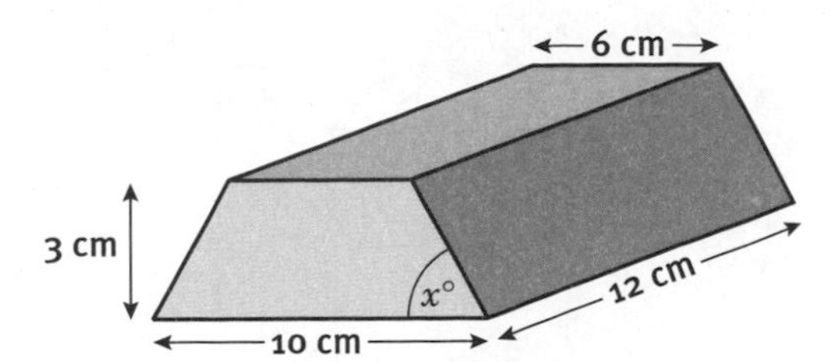

(a) Calculate the volume of the bar.

....

.... **[4]**

(b) Given that the density of lead is 11.35 g/cm^3, calculate the mass of the bar.
Give your answer in kg correct to 3 significant figures.

....

.... **[1]**

(c) Calculate the area of the sloping face. Give your answer correct to 3 significant figures.

....

.... **[3]**

(d) Calculate the angle $x°$ which the sloping face makes with the base.

....

.... **[3]**

(Total 11 marks)

7 **(a)** Factorise

(i) $3x^2 + 15x$

.... **[2]**

(ii) $x^2 - 5x - 36$

....

.... **[2]**

(b) Simplify

$$\frac{m^3p^4 \times m^2p^5}{m^4p^6}$$

....

.... **[2]**

(Total 6 marks)

[turn over

8 **(a)** On the grid below, using values of x from 0 to 6, draw the graphs of $2x + 3y = 12$ and $3x - 2y = 3$.

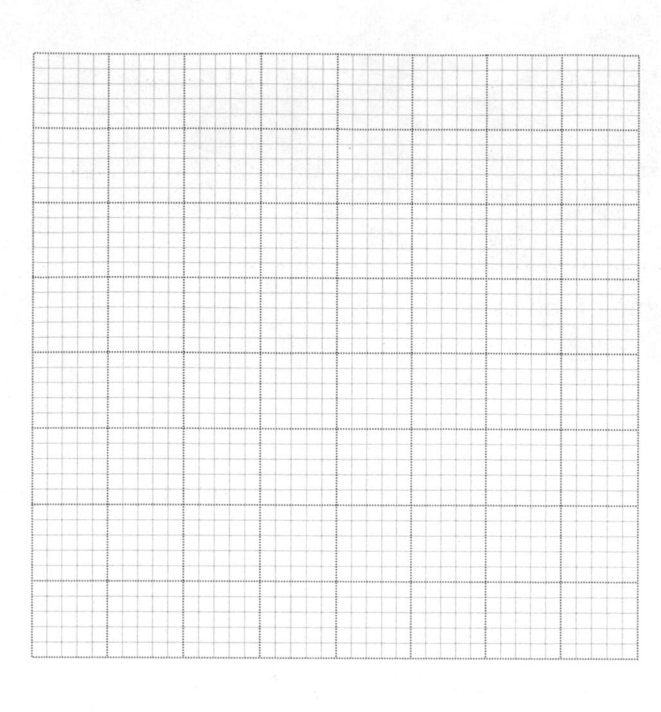

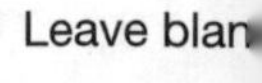

[5]

(b) Use your graph to solve the simultaneous equations

$$2x + 3y = 12$$
$$3x - 2y = 3$$

$x =$ ....; $y =$ .... **[1]**

(Total 6 marks)

9 In this question $a = 2 - 3\sqrt{2}$.

(a) Express a^2 in terms of $\sqrt{2}$ simplifying your answer as much as possible.

....

.... **[2]**

(b) Hence show that $x = a$ is one solution of the equation

$$x^2 - 4x - 14 = 0$$

....

....

.... **[1]**

(Total 3 marks)

10 The diagram below shows a rectangle, A, length 11 cm and width y cm, and a square, S, side x cm.

11 cm
A
y cm

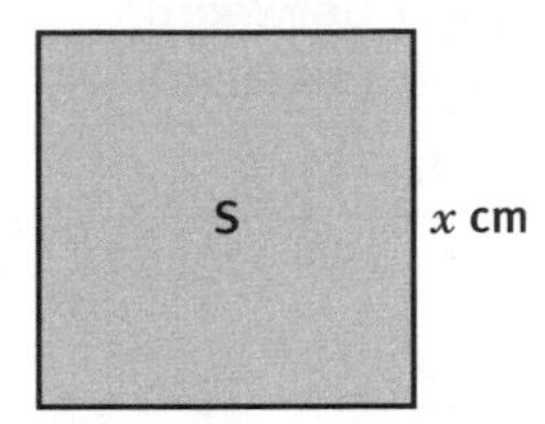

(a) Write down an expression for the perimeter of

(i) the rectangle

[1]

(ii) the square.

[1]

The perimeter of the square and the rectangle are equal.

(b) Use this fact to write down an expression for y in terms of x.

[2]

The area of the square is $4\,\text{cm}^2$ more than the area of the rectangle.

(c) **(i)** Write down an equation satisfied by x and show that it simplifies to

$$x^2 - 22x + 117 = 0$$

[2]

(ii) Solve this equation, giving the two possible values of x.

[3]

(Total 9 marks)

Leave blank

11 The time between 150 successive aircraft landing at a busy airport, during the period 0800–1400 one day, is given in the table. For example, on 24 occasions an aircraft arrived within 60 seconds of the aircraft in front of it.

Leave blank

Time t between aircraft (seconds)	Number of aircraft i.e. frequency
$0 \leqslant t < 60$	24
$60 \leqslant t < 120$	34
$120 \leqslant t < 180$	27
$180 \leqslant t < 300$	38
$300 \leqslant t < 420$	27

(a) Estimate the mean time between arrivals.

....

....

....

.... **[3]**

(b) On the grid below draw a histogram to show this data.

....

....

....

.... **[4]**

(Total 7 marks)

Leave blank

12 The pressure of the water, P, at any point below the surface of the sea varies as the depth, D, of that point below the surface.

The pressure is 400 newtons/cm^2 at a depth of 6 m.

Find the pressure at a depth of 20 m.

Give your answer correct to the nearest whole number.

....

....

....

.... **[3]**

(Total 3 marks)

13 A solid metal cone of height 100 cm and base radius 70 cm is melted down and recast as two identical solid cylinders of length 60 cm.

Calculate the radius of the cylinders.

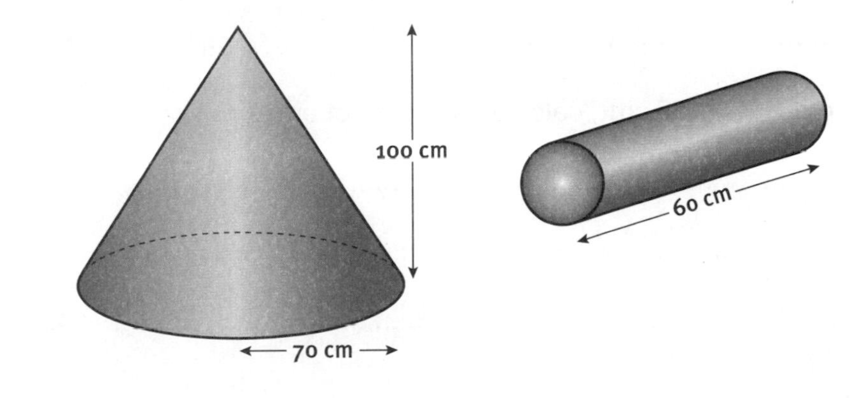

....

....

....

.... **[5]**

(Total 5 marks)

14 In the triangle ABC, AB = 10 cm, AC = 15 cm and angle BAC = x°.

Leave blank

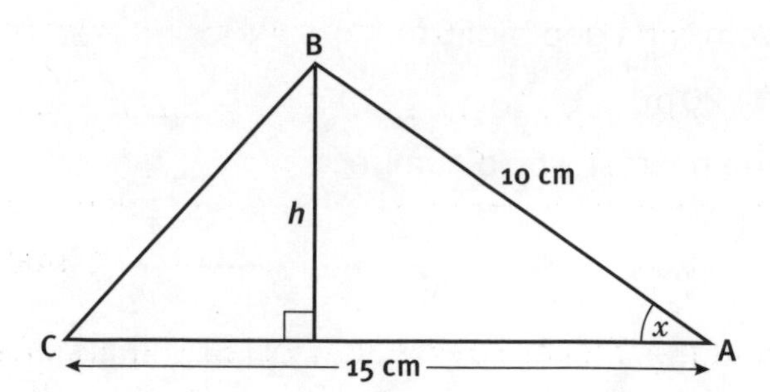

(a) Write down an expression for the height h in terms of x.

....

.... **[1]**

(b) Show that the area of triangle ABC is 75 sin x cm^2.

....

....

.... **[1]**

(c) If the area of triangle ABC is 50 cm^2, calculate the size of angle x.

....

....

.... **[2]**

The area of this obtuse-angled triangle PQR is also 50 cm^2.

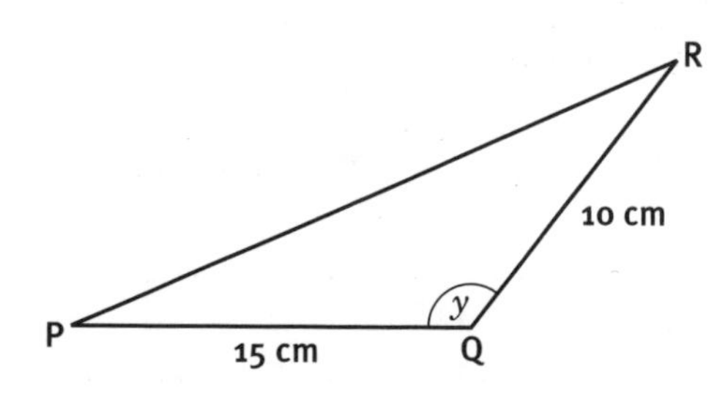

(d) Calculate the size of angle y.

....

....

.... **[1]**

Leave blank

(e) Calculate the length of PR.

....

....

.... **[4]**

(Total 9 marks)

15 The dimensions of the cuboid in the diagram below are accurate to the nearest cm.

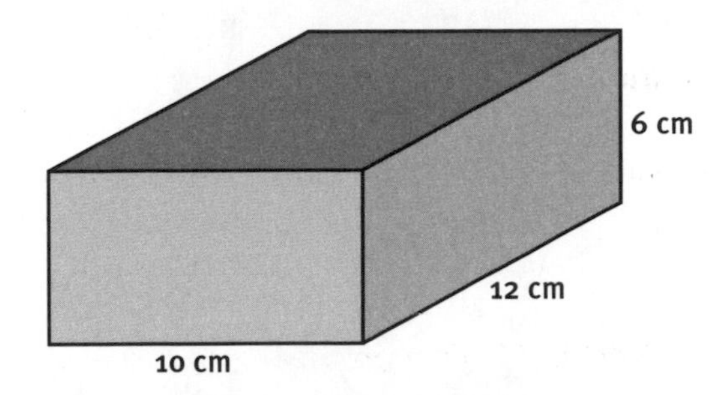

What is the difference between the greatest true volume and the least true volume? Give your answer correct to two decimal places.

....

....

....

.... **[3]**

(Total 3 marks)

16 Use algebra to solve the equation:

$$\frac{4}{x+3} - \frac{3}{x+4} = 1$$

....

....

....

....

....

.... **[7]**

(Total 7 marks)

[turn over

Answers to mock examination Paper 1

1 (a) $2 \times 3 \times 5$ **[2]**
(b) (i) 270 **[1]**
(ii) 6 **[1]**

EXAMINER'S TIP

You should know Lowest Common Multiple (LCM) and Highest Common Factor (HCF).

2 (a) $5\frac{11}{12}$ **[3]**
(b) $\frac{5}{12}$ **[3]**

EXAMINER'S TIP

Don't forget the common denominator.

3 £200 **[3]**

EXAMINER'S TIP

Use the multiplier.
Divide by 0.8.

4 (a) $c = y - mx$ **[1]**
(b) $c = \pm\sqrt{\dfrac{e}{m}}$ **[2]**
(c) $c = \dfrac{4}{b - 2a}$ **[2]**

EXAMINER'S TIP

*In **(c)** collect the c's first.*

5 (a) $EC = 3 \times 3 - 3 = 6$ **[2]**
(b) $AD = \frac{1}{2}DB = 1\frac{1}{2}$ **[2]**

EXAMINER'S TIP

Triangles ADE and ABC are similar so their sides are in proportion.

6 (a)

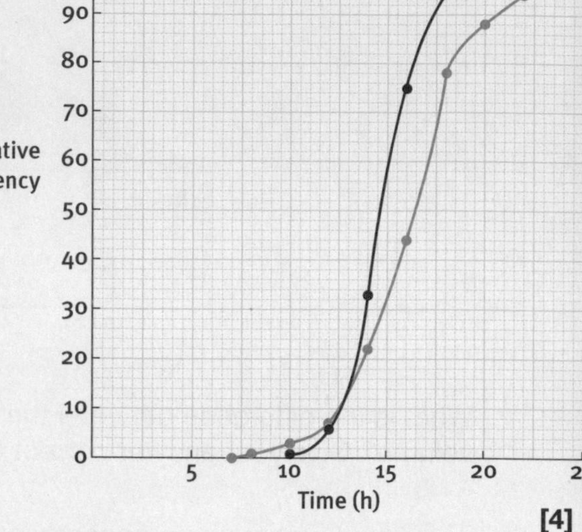

[4]

(b) *Britelite* batteries

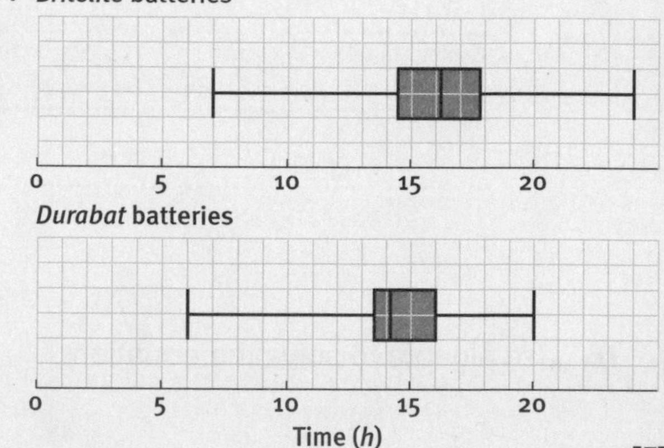

[5]

(c) Durabat last longer although a few failed early. **[1]**

EXAMINER'S TIP

Be careful to show which plot is which.

7 (a) $(2x - 1)(x + 3)$ **[2]**
(b) (i) $x = \frac{1}{2}, -3$ **[1]**
(ii) $-3 \leqslant x \leqslant \frac{1}{2}$ **[2]**

8 (a) $x = 16$ **[2]**
(b) $x = 1\frac{1}{2}, y = -2$ **[4]**

EXAMINER'S TIP

*You need to operate on both equations in **(b)** in order to eliminate one variable*

9 **(a)** AD = CB (opposite sides of a parallelogram)
DC = BA (opposite sides of a parallelogram)
AC common
Triangles congruent (SSS) **[3]**

(b) Rotation, half-turn about mid-point of AC **[3]**

EXAMINER'S TIP

To describe a transformation be sure to include all the details.

10 $0.6 \times 0.2 + 0.4 \times 0.3 = 0.24$ **[4]**

EXAMINER'S TIP

You may find it helpful to draw a tree diagram.

11 **(a)** $10\sqrt{5}$ **[1]**

(b) $5 + \sqrt{5}$ **[1]**

(c) $-1\frac{1}{2} - \frac{1}{2}\sqrt{5}$ **[3]**

EXAMINER'S TIP

*Be careful with the signs in **(c)**.*

12 **(a)** Read at $y = 0$ (1.8 or 1.9) **[1]**

(b) Read at $y = -1$ (0, 1 repeated) **[2]**

(c) Read at $y = 1 - x$ (1.5) **[3]**

(d) Read at $y = x - 2$ (−0.6, 1.1, 1.6) **[3]**

(e) Yes, one more off the diagram (left) as cubic graph gets steeper. **[2]**

EXAMINER'S TIP

Draw the required lines ($y = -1, y = 1 - x, y = x - 2$) on the graph.

13 **(a) (i)** $-\dfrac{x+2}{6}$ **[2]**

(ii) $\dfrac{x+2}{x(2x+1)}$ **[4]**

(b) (i) $\dfrac{x}{x+1}$ **[3]**

(ii) $\dfrac{b^2}{2a}$ **[3]**

EXAMINER'S TIP

Show all the steps of your working.

14 **(a)** Angle ABC = 62° (angle at circumference half angle at centre)
Angle ACT = 62° (angle in alternate segment)
(or use isosceles triangle AOC) **[3]**

(b) B = 62°, A = x, C = 90° − x + 28° = 118° − x. **[3]**

EXAMINER'S TIP

Make your reasons clear to earn the marks.

15 **(a)** False since volume scale factor is the cube of the linear scale factor. **[2]**

(b) False since cube root of $\frac{1}{2}$ is greater than $\frac{1}{2}$. **[2]**

(c) True, $\frac{32}{25} = 1.28$ and $1.28^3 \approx 2$ **[2]**

EXAMINER'S TIP

This is about using the volume scale factor and giving reasons.

16 **(a)**

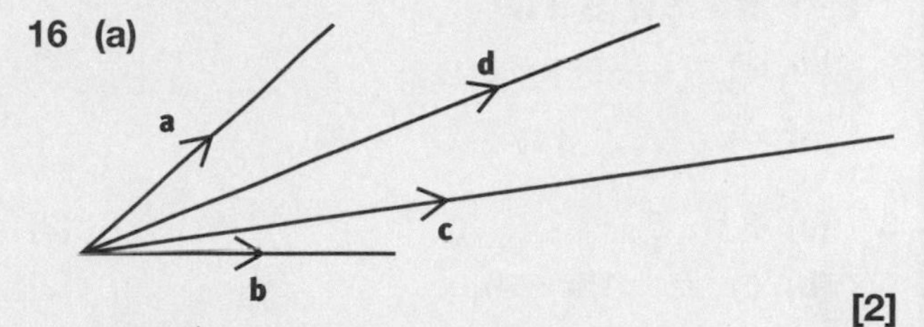

[2]

(b) $\overrightarrow{AD} = 2\mathbf{a} + \mathbf{b} - \mathbf{a} = \mathbf{a} + \mathbf{b}$ **[2]**

(c) $\overrightarrow{BC} = \mathbf{a} + 2\mathbf{b} - \mathbf{b} = \mathbf{a} + \mathbf{b}$
$\overrightarrow{AD} = \overrightarrow{BC}$, opposite sides equal and parallel **[3]**

EXAMINER'S TIP

It will still work if you take the letters in the wrong order, $\overrightarrow{CB} = -(a + b)$

Answers to mock examination Paper 1

Answers to mock examination Paper 2

1 0.68 [4]

EXAMINER'S TIP

You will need to calculate x to 3 decimal places to decide whether to round up or down.

2 (a) (i) 458.59 [2]

(ii) 0.80 [2]

(b) $3.87 \times 10^{-1} : 1$ [2]

EXAMINER'S TIP

*In **(a) (i)** use brackets.*

3 (a) $3m - 8$ [2]

(b) (i) $(x + 1)(x - 4)$ [2]

(ii) $x = -1; x = 4$ [1]

EXAMINER'S TIP

*Use the answer from **(b) (i)** to write down the answer to **(b) (ii)**.*

4 (a) £748 [2]

(b) £68.57 [2]

(c) £1304 [3]

EXAMINER'S TIP

*In **(a)** do not add the percentage reductions together – treat them as sequential.*

5 (a) $P(WW) = \frac{4}{10} \times \frac{2}{8} = \frac{1}{10}$

$P(BB) = \frac{3}{10} \times \frac{2}{8} = \frac{3}{40}$

$P(GG) = \frac{3}{10} \times \frac{4}{8} = \frac{3}{20}$

$P(\text{same colour}) = \frac{1}{10} + \frac{3}{40} + \frac{3}{20} = \frac{13}{40}$ [5]

(b) must be a white shirt,

$P(W) = \frac{6}{10} \times \frac{5}{9} \times \frac{4}{8} = \frac{1}{6}$ or $0.16\dot{6}$ [4]

EXAMINER'S TIP

*In **(a)** be systematic.*

6 (a) $288\,\text{cm}^3$ [4]

(b) 3.27 kg [1]

(c) $43.3\,\text{cm}^2$ [3]

(d) 56.3° [3]

EXAMINER'S TIP

*In **(b)** be careful with units.*

7 (a) (i) $3x(x + 5)$ [2]

(ii) $(x - 9)(x + 4)$ [2]

(b) mp^3 [2]

EXAMINER'S TIP

*In **(b)** remember the rules about adding and subtracting indices*

8 (a)

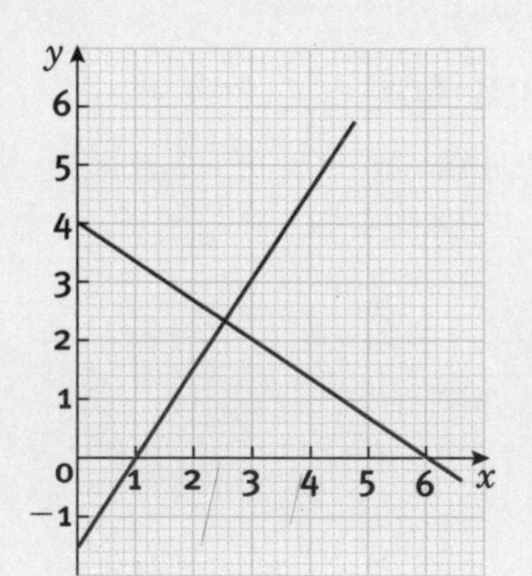

[5]

(b) $x = 2.5, y = 2.3$ [1]

EXAMINER'S TIP

*In **(a)** calculate the values of y when $x = 0$ before drawing the axes.*

9 (a) $2(11 - 6\sqrt{2})$ [2]

(b) $22 - 12\sqrt{2} - 4(2 - 3\sqrt{2}) - 14$

$= 22 - 8 - 14 - 12\sqrt{2} + 12\sqrt{2}$ [1]

EXAMINER'S TIP

*In **(b)** substitute a and a^2.*

10 (a) (i) $2y + 22$ **[1]**

(ii) $4x$ **[1]**

(b) $y = 2x - 11$ **[2]**

(c) (i) $x^2 - 4 = 11(2x - 11)$ **[2]**

(ii) 13, 9 **[3]**

EXAMINER'S TIP

*In **(c)** remember to subtract 4 from x^2 to equate with the area of the rectangle.*

11 (a) 177.8 seconds **[3]**

(b) Frequency densities are 0.4, 0.57, 0.45, 0.32 and 0.23

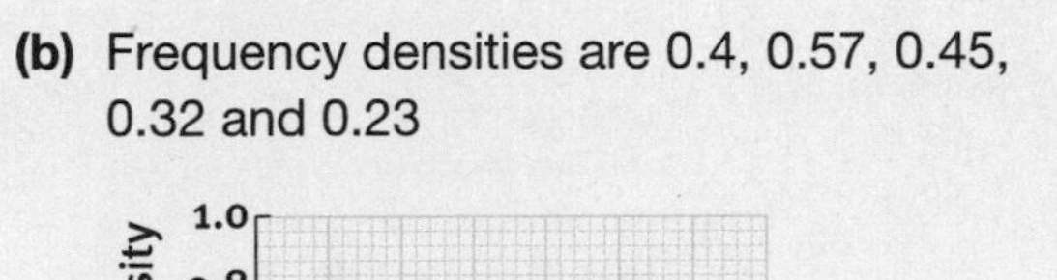

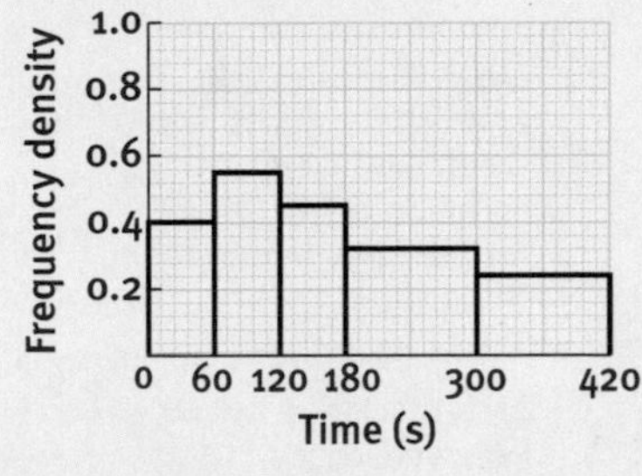

[4]

EXAMINER'S TIP

Remember to use mid-interval values and calculate frequency density.

12 1333 N/cm^2 **[3]**

EXAMINER'S TIP

Because the question is concerned with proportionality you do not have to convert to cm or to m.

13 36.9 cm **[5]**

EXAMINER'S TIP

Remember the question is concerned with 2 cylinders.

14 (a) $10 \sin x$ **[1]**

(b) $A = \frac{1}{2}$ base $\times$ height

$= \frac{1}{2} \times 15 \times 10 \sin x$

$= 75 \sin x$ **[1]**

(c) 41.8° **[2]**

(d) 138.2° **[1]**

(e) 23.4 cm **[4]**

EXAMINER'S TIP

*In **(d)** use the formula area of a triangle $= \frac{1}{2} ab \sin C$.*

15 252.25 cm^3 **[3]**

EXAMINER'S TIP

Remember the difference is $(12.5 \times 10.5 \times 6.5) - (11.5 \times 9.5 \times 5.5)$

16 -5, or -1 **[7]**

EXAMINER'S TIP

Multiply through by $(x + 3)(x + 4)$ giving $4(x + 4) - 3(x + 3) = (x + 3)(x + 4)$ and solve this.

Index